DRAWING COASTLINES

EXPERTISE

CULTURES AND
TECHNOLOGIES
OF KNOWLEDGE

EDITED BY DOMINIC BOYER

DRAWING COASTLINES

Climate Anxieties and the Visual
Reinvention of Mumbai's Shore

V. Chitra

CORNELL UNIVERSITY PRESS   **ITHACA AND LONDON**

First published 2024 by Cornell University Press

Printed in the United States of America

Library of Congress Cataloging-in-Publication Data

Names: Chitra, V. 1982– author.
Title: Drawing coastlines : climate anxieties and the visual reinvention of Mumbai's shore / V. Chitra.
Description: Ithaca : Cornell University Press, 2024. | Includes bibliographical references and index.
Identifiers: LCCN 2024008882 (print) | LCCN 2024008883 (ebook) | ISBN 9781501777967 (paperback) | ISBN 9781501777974 (epub) | ISBN 9781501777981 (pdf)
Subjects: LCSH: Nature—Effect of human beings on—India—Mumbai. | Coasts—Social aspects—India—Mumbai. | Charts, diagrams, etc.—Social aspects—India—Mumbai. | Coastal mapping—Social aspects—India—Mumbai. | Climatic changes—Social aspects—India—Mumbai.
Classification: LCC GF75. C475 2024 (print) | LCC GF75 (ebook) | DDC 333.91/70954792—dc23/eng/20240514
LC record available at https://lccn.loc.gov/2024008882
LC ebook record available at https://lccn.loc.gov/2024008883

Contents

Acknowledgments

It has taken me almost a decade to write and draw this book, and it would not have been possible without all the support I have had the privilege of receiving.

Deborah Poole and Anand Pandian have both, in their own ways, shaped me as a scholar and artist. Their offices were safe spaces where I could fail over and over again. I can only hope to make such space for others. I am also thankful to Naveeda Khan whose keen insights shined clear paths through many a fog of ideas. Stuart W. Leslie went above and beyond in his role as a committee member and continues to be an important and kind mentor. Both Erica Schoenberger and Rebecca Brown helped tie together important sections of this book.

I am extremely grateful to the team at Cornell University Press—Dominic Boyer, Jim Lance, Marshall R. Hopkins, Clare Jones, Ellen Labbate, Karen M. Laun, Kimberly Maselli Schmelzinger, Jack Rummel, and others for their support. Dominic's and Jim's enthusiasm boosted my confidence—perhaps this can actually work, I thought. To the two anonymous reviewers who provided helpful comments and encouragement—thank you, it made a huge difference to me, and I hope you can see your important contribution in the book's final form.

My extended host family at Malvani Koliwada, and the team that worked on their boat, gave me an immense amount of attention and care. I am sorry I cannot thank you by name. Similarly, many thanks are owed to the survey team from Anna University, to the surveyors from the P (North) ward office, the personnel at the Municipal Corporation of Greater Mumbai, Marine Life of Mumbai team, and the fishers at Marve Beach and Malad Station market. I owe special thanks to Mr. Narendra Pagare and Ms. Neeta Gaikwad at the Municipal Corporation of Greater Mumbai who have the tough job of making enormously complex urban infrastructures run. I am grateful to T. I. Eldho, Arun Inamdar, Dipanjali Majumdar, and R. Nagarajan for the patience with which they explained technical processes and their ecological outcomes. I greatly benefited from Hrishikesh Samant's deep knowledge of the region's geological makeup and history. Thanks are also owed to Amita Bhide, Siddharth Chakravarty, Gopal M. S, Hussein Indorewala, Abhishek Jamlabad, Lalitha Kamath, Hareshwar Koli and family, Rajesh Mangela, D. Parthasarthy, Pradip Patade, Aslam Saiyad, Makarand Salunke, Madhuri Shivkar, Raj Hans Tapke, Aravind Unni, and Shweta Wagh for taking time to answer my questions and pointing me in the right direction when I asked the wrong questions. Last, I owe a debt of gratitude to Akshay Sawant and Sheetal Gokhale, whose help made my research better.

My graduate research and writing was made possible by grants from the Environment, Energy, Sustainability, and Health Institute and the Center for Advanced Media Studies at Johns Hopkins University. The National University of Singapore's (NUS) faculty start-up grant supported my second phase of research, and the Faculty of Arts and Social Science's book grant aided its publication. The Department of Sociology and Anthropology made this book possible by moving my teaching load, giving me time to write when I needed it. A residential fellowship at the Akademie Schloss Solitude allowed me to shut off the world and draw and edit in the most magical environment.

Three Mumbai-based institutions—Kamla Raheja Vidyanidhi Institute for Architecture (KRVIA), the Indian Institute of Technology Bombay, and more recently, School of Environment and Architecture have generously supported my research and provided space, affiliation, and a scholarly community. A postdoctoral fellowship at the Laxmi Mittal South Asia Institute at Harvard University made it possible for me to take stock of the larger project. Thanks are also owed to the archivists at the Harvard Map Collection, the Mumbai Suburban Land Records Office and Fisheries Department, the Maharashtra State Archives, and the Mumbai Metropolitan Regional Development Authority's library.

Parts of this book have appeared as articles in *American Ethnologist* and *Urban Studies*. A chapter for a volume on housing in India titled "Identification, Materiality and Housing Transformations in Mumbai" helped coalesce my understanding of the coastal policy and its implications. Sets of comics appeared in an *Anthro News* feature on graphic ethnographies, and an online exhibition on "Illustrating Anthropology." Thank you to the editorial teams, to the anonymous scholars who took the time to review my work, and to the editors and curators I worked with, including Jennifer Cearns, Karen Coelho, Alexandra Frankel, Laura Haapio-Kirk, Michael Hathaway, Natalie Konopinski, Stacy Leigh Pigg, Urmi Sengupta, and Annapurna Shaw. I am very grateful to scholars who invited me and provided feedback at conferences and workshops—Nikhil Anand, Sai Balakrishnan, Lisa Björkman, Christiane Brosius, Dana Burton, Shubra Gururani, Theo Hilton, Arun Inamdar, Jane Jacobs, Lalitha Kamath, Shweta Krishnan, Ratoola Kundu, Anthony Medrano, Sheehan Thomas Moore, Rahul Mehrotra, Deborah Shamoon, K. Sivaramakrishnan, and Anand Vaidya. Kausik Mukhopadhyay and the faculty at the Industrial Design Center gave me an excellent education in the visual arts.

I am ever thankful to my colleagues and friends at my former academic home, NUS's Department of Sociology and Anthropology, particularly Noorman Abdullah, Ho Kong Chong, Daniel Goh, Kelvin Low, Feng Qiushi, Rajesh

Rai, and especially, Vineeta Sinha, who has been a great support over the years. The anthro reading group at NUS was a good testing ground for the first chapter. Jennifer Estes, Zach Howlett, Ting Hui Lau, Canay Özden-Schilling, Tom Özden-Schilling, Elliott Prasse-Freeman, and Mu Zheng have built a wonderfully supportive environment. The staff team—Mr. Fuqaimi, Haseena Bte Hashim, Jane Ong, K.S. Rajamani, and Vickneswari Savarimuthu—smoothed many an administrative impediment. At my new home at the Australian National University, I continue to find enthusiasm for ethnographic experiments.

I owe thanks to Sidharthan Maunaguru for encouraging me to chase my project all the way to Singapore. Sid and Anojaa Karunananthan's place was my first home there. Alphreda Antony, Ameya Athavankar, Sai Balakrishnan, Kara Gionfriddo, Rupali Gupte, Kristina Hallez, Aditi Jha, George Jose, Ranjit Kandalgaonkar, Prasad Khanolkar, Ateya Khorakiwala, Namrata Kapoor, Avijit Mukul Kishore, Arjita Mathur, Kate O'Brien, Apurva Parikh, Lubaina Rangwala, Savita Rauthan, Surabhi Sharma, Prasad Shetty, Rohan Shivkumar, Sonal Sundararajan, Sweta Suryanarayanan, and Abhijit Visaria—your love buoys me; thank you for always being there for me. In Baltimore, Hannah Friedman and Emma McGlennen were early readers and my home away from home. I am deeply grateful to Kellan Anfinson and Derek Denman for their staunch camaraderie, for taking the time to read chapter drafts, and for spirited Zoom-beer-hours. Itty Abraham, Chandan Bose, Namita Dharia, Rohan D'Souza, Colin Hoag, Hansika Kapoor, Shweta Krishnan, Anirudh Paul, Nayanika Mathur, Gayatri Nair, and Tarangini Sriraman generously provided their thoughts, friendship, and wise counsel. Ben Cowburn edited and proofread mangled drafts and helped them make sense. Ratheesh Radhakrishanan very kindly let me use his office for a summer—thank you, I got so much writing done.

So many people have lived with this book, and with me living this book. I am grateful for the love my family in Germany—Kristof, Markus, Ursula Schulze, Julia Zinke, and the late Dieter Schulze—gave me. Sneha Annavarapu, Shawn Chong, Kevin Goldstein, Shailey Hingorani, Rohan Mukherjee, and Kriti Vikram were always there for me, with food/drink/medicine/dance music/meme, depending on the time of day and circumstance. With Nienke Boer's friendship, I was able to overcome the inevitable teething troubles and imagine the book. Timothy McDonald's love, humor, and gentleness makes every day lighter and happier—I look forward to our adventures. Tapsi Mathur, person and scholar extraordinaire, thank you for your sparkling wit, and for being ever ready to dream up plotlines for alternate lives. Thank you to my canine and feline friends—Charles, Dio, Foxy, Frida, and even Puffin (despite her rejection)—for artfully demonstrating the value of living in the present. My parents, N. Venkataramani and V. Kamala, are always supportive of their children's dreams and make it possible for us to be intrepid.

This book is dedicated to two creatures who hold me in their hearts: my sister Kavita Venkataramani, thank you for being a great sibling, and for introducing me to the beauty of fig-wasp worlds. And to Bubbles—thank you for being my companion for all the lines in this book, snoozing softly with one eye open for treats or monsters, both always equally probable.

Abbreviations

CRZ	Coastal Regulatory Zone
CZMA	Coastal Zone Management Authority
CZMP	Coastal Zone Management Plans
DP	Development Plan
GIS	Geographic Information Systems
GPS	Global Positioning Systems
HTL	high tide line
ICSF	International Collective in Support of Fishworkers
KRVIA	Kamla Raheja Vidyanidhi Institute of Architecture
LTL	low tide line
MCGM	Municipal Corporation of Greater Mumbai
MMKS	Maharashtra Macchimar Kruti Samiti
MMRDA	Mumbai Metropolitan Regional Development Authority
MOEFCC	Ministry of Environment, Forests, and Climate Change
NFF	National Fishworkers Forum
SoI	Survey of India
TISS	Tata Institute of Social Sciences
UDRI	Urban Design Research Institute
YUVA	Youth for Unity and Voluntary Action

Note on Names, Transliteration, and Comics

All names are pseudonyms unless noted otherwise. I have used the original place names where the studies are publicly available. The research was approved by the Johns Hopkins University's Internal Review Board (granted exemption, HIRB 2011079) and later, by the National University of Singapore's Internal Review Board (LS-17–093E).

The book shifts between comics and text, and they are to be read in continuation with one another, and not as separate parts. The conversations in this book move between English, Hindi, and Marathi. I have simplified the transliteration, sticking close to the sounds to capture the texture and tone of the conversation. I do not italicize non-English words in the comics and only italicize them once in the text, when they first appear. The sounds of things are often communicated using the Devanagari script. I saw this in translated manga, which many a time retained the original Japanese given the difficulty of translating sound effects that are a part of the art, leaving the readers to rely on visual onomatopoeia—I wanted to play with this. Citations in the comics are included in their respective notes. The fonts used in comics are DigitalStrip BB by Blambot, SF Action Man by ShyFoundry, and Bohemian Typewriter by Lukas Krakoa.

Any errors in the book are mine alone.

SIGHTLINES

IN 2011, THE INDIAN GOVERNMENT RELEASED A NEW VERSION OF ITS COASTAL POLICY. ITS RULES SET BOUNDARIES, DISTRIBUTE RIGHTS, AND DESIGNATE RESOURCE, CONSERVATION, AND HERITAGE VALUE TO COASTAL ELEMENTS.

THIS POLICY DECIDES WHAT THE COAST IS AS A TERRITORIAL, SOCIOPOLITICAL, AND MATERIAL ENTITY.

KUMB

COASTAL CONTESTATIONS

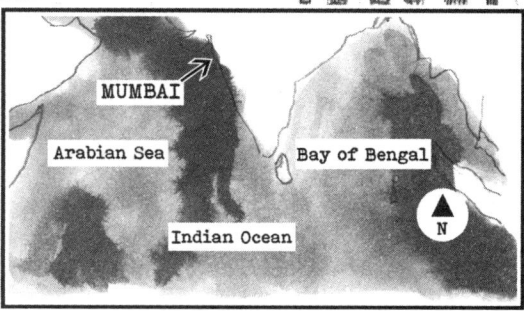

MUMBAI

Arabian Sea

Bay of Bengal

Indian Ocean

N

IN THE COASTAL CITY OF MUMBAI, WHERE MORE THAN 20 MILLION PEOPLE RESIDE, THE TASK OF MAKING THESE PLANS FELL TO THE MUNICIPAL CORPORATION OF GREATER MUMBAI (MCGM). THE MCGM CONTRACTED A TEAM OF REMOTE SENSING AND MAPPING SPECIALISTS FROM ANNA UNIVERSITY, CHENNAI, FOR THE JOB.

IN 2012, I FOLLOWED THE TEAM AS THEY SURVEYED MUMBAI'S SHORE FOR THE COASTAL PLANS.

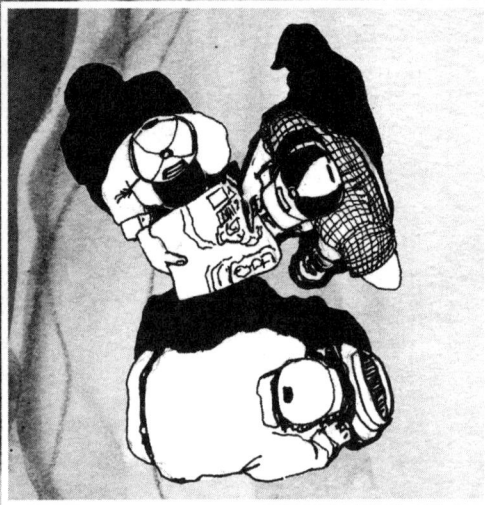

THE TEAM WORKED WITH SPEED AND EFFICIENCY TO FINISH THEIR TASK ON TIME. YET, THEY WOULD OFTEN PAUSE TO ANSWER MY QUESTIONS AND EXPLAIN SYSTEMS AND PROCESSES.

ONE DAY, WE WERE STANDING AROUND TAKING SURVEY READINGS AT A SEWAGE TREATMENT PLANT IN THE MANGROVES.

ANBU, A GRADUATE STUDENT WHO WAS WORKING ON THE PROJECT, AND I WERE TALKING ABOUT THE VALUE OF COASTAL PLANS AND WHERE THEY FIT IN THE HISTORIES OF SURVEYING AND MAPPING. ANBU WAS STARING OUT INTO THE OPEN FILTER BEDS, WHERE DARK SLUDGY FLUID GATHERED AIR.

POINTING IN THE GENERAL DIRECTION OF THE SEWAGE, HE SAID THAT WITHOUT THE PLAN THERE WAS NO WAY OF KNOWING WHERE THE COAST BEGAN OR ENDED OR WHERE ONE COULD BUILD OR NOT BUILD THINGS.

THE COAST IS A GEOGRAPHIC ENTITY, HE REASONED, SO HOW COULD IT BE PLANNED WITHOUT MAPPING THE THINGS IN IT? IF THE POLICY DEPENDED ON LOCATING THINGS AND MARKING ZONES, IT COULD NOT EXIST WITHOUT DRAWINGS THAT DID THAT WORK.
THIS ONE IS GOOD.
WHERE NEXT?
OTHER SIDE?

THEY ARE NOT MERE REFLECTIONS OR SIMPLY MEDIUMS FOR FINDING CORRESPONDENCE BETWEEN THE POLICY AND THE LANDSCAPE. NOR DO THEY REMAIN IN THE REALM OF THE VIRTUAL OR IMAGINED.

TECHNICAL DRAWINGS AND THEIR ATTENDANT CATEGORIES, LABELS, AND LINES SET THE CONDITIONS OF THE COAST.

COASTAL PLANS ARE GOOD EXAMPLES OF TECHNOSCIENTIFIC IMAGES IN ACTION. THEY MANIFEST AS THE COAST, DRAWING ITS REALITIES AND FUTURES INTO BEING.

This book is about how technoscientific drawings create coasts. It dwells on the lives of these images from Mumbai and in it, I follow surveyors, fishing communities, planners, amateur meteorologists, marine life advocates, and their visual projects. The book's ethnographic encounters and visual analyses show how drawings transform the coast into an object of management, infrastructure, fisheries, an urban edge, and an ecology.

Because most drawings are produced in the service of policies that treat the environment as a resource, they take a short-term view of the coast. They ignore its fluid liminalities, fixing dynamic terraqueous worlds in developmental terms. Taken as representations of objective and apolitical scientific enterprise, they perpetuate cycles of colonization. At the same time, scientific images are vital to environmental action, participatory politics, and indigenous claims, and they are the objects around which climate anxieties coalesce. Technical images such as plans, surveys, graphs, and infographics remake human-nonhuman relationships, arrange urban politics, and materialize landscapes in complex and contradictory ways. These incongruities are opportunities to envisage new ways of drawing—and thus, creating coasts—that move away from solution-oriented frameworks and acknowledge the multispecies worlds and material continuities at the shoreline.

Coastlines are good sites to think about the relation between technical images, landscape, and climate change because they are where the effects of the crisis are both acutely felt and hypervisible. Technical images abound at the shore; from maps that forecast coastal change with sea-level rise to before–after pictures of cyclones, they are sites where a range of visual technologies converge to measure, project, estimate, and plan for the destruction that is now inevitable. These images are not singular objects nor is their power solely in the information they carry. They are often a coming together of many technologies, databases, and practices. They exist in relation to other images and image histories. Their power is inseparable from their material conditions—whether they exist on paper, as manipulable vector lines on a screen, as barely legible photocopies, or fuzzy lines on a cellphone image shared for a brief instance.

I use the term *drawing* to describe the plurality of images and the many ways in which they become the shore. Drawing refers to different things, actions, and possibilities. It refers to the object itself, and it refers to the act of drawing—such as when a surveyor makes a line on a page or a screen to mark a location or feature. It also refers to the act of drawing together different images and datasets to make an image, such as when surveyors compile different images to make coastal plans. For example, while talking about the process of creating the coastal plans, Anbu explained that the Coastal Zone Management Plans (CZMPs) were multi-layered drawings of which the ground surveys were one part. Information about individual property and plot boundaries was also needed to create a granular scale at which the plans were supposed to work. To that, the survey team would add information such as forest and district boundaries. Moreover, if the government decided to add a hazard line to prepare for tsunamis and sea-level rise, the map creators would need other data, including tide tables, climate change projections, and elevation. Thus, images like the CZMP draw together a variety of historical and contemporary visual datasets that utilize different technologies. They have to be updated when things change on the ground or as policies are revised.

Apart from the multiple elements they draw together, technoscientific images also pull together communities, technologies, and institutions. These include the scientific teams that create the images, the institutions and bodies that set the visual parameters, the technologies and practices used to make them (ranging from pencils and logbooks to satellites, and mapping and imaging software), and communities who see and respond to them with drawings of their own. This was evident in the way technical drawings served as a canvas for public debates on the changing coastal policy. Last, to draw also conveys an action or object that has duration rather than a representation that freezes a likeness in time. Many of the images I encountered envisioned coastal change in relation to developmental goals, working with plans that operated along decadal cycles. Thus, in using the term *drawing*, I am interested in how images act and pull together landscapes, things, institutions, and communities, and in the way they cast futures, which unfold along multiple temporal horizons.[1]

I began this project in 2011, the same year that the Indian government proposed revisions to the Coastal Regulatory Zone (CRZ) policy. It was also the year that the Municipal Corporation of Greater Mumbai (MCGM) redoubled its effort to revise Mumbai's Development Plan (DP), which had reached the end of its twenty-year cycle. The DP is a document that guides urban transformation, including land use, building construction, green spaces, amenities, industry, and infrastructure. Mumbai's indigenous fisher community, the Kolis, found themselves in the crosshairs of these projects, both of which opened coastal land for development—prime real estate in a land scarce city. To secure their right to the coast, the Kolis forged strategic alliances with planners, urban activists, and architects, and initiated projects to map their own coastal visions. Technical images became the material objects around which fishers, planners, and surveyors organized themselves as publics, despite their differing histories and relationships with mapping and planning. Their drawings constructed divergent claims, even as they shared much in terms of visual language and political rhetoric with each other and with the state's plans. Each image called a different coast and different coastal futures into being, though they were all animated by similar ideas of anthropocentric environmental sustainability. The immediate problem of securing rights and avoiding mass displacement took on a sense of urgency that overshadowed the fisher community's environmental concerns, particularly of those engaged in small-scale fishing. The anxieties about coastal pollution, changing weather patterns, and depleting catch were always lurking at the edges, but exigent problems demanded a different temporal focus, and this is evident in the plans and maps drawn at this time.

Technical images—particularly cartographic plans—have long shaped spatial politics, land rights, and belonging in Mumbai in intricate ways.[2] They are largely taken as mediators of context-specific politics, tied to the places they map and theorized in relation to questions of rights and resource distribution. These questions are no doubt important for the communities I interacted with, but these images also exist in a global visual economy that molds the experience of climate change.[3] This period also marks an important time in the history of the Internet in India, as GIS technology, digital governance, online image repositories, and social channels gained popularity. As virtual technologies became easier to access, they spawned communities that engaged the environment through visual technologies that were, until recently, confined to state institutions and domains of expertise.[4]

These images—of the sort that make up the CZMPs—are inseparable from histories of colonization (Jay 2014). They are material objects that articulate empire and imperial subjectivities (Ramaswamy 2014). In India, as in other parts of the colonial empire, the vertiginous view of cartographic images allowed British officers to empty the landscape of its rich ecological relations and impose a governmental order that fueled the global colonial economy (Bhattacharyya 2018; Edney 1997).[5] In present times, technical images continue to territorialize—surveys, plans, graphs often fate landscapes, animals, plants, and humans to catastrophic futures. They implement management programs that separate the environment from human domains, rendering it as resource to sustain transnational flows of capital (Carruth and Marzec 2014; Marzec 2015). Yet, they are also images that are fundamental to climate politics: graphs of rising temperature and global supply chain maps confront their viewers with how they consume the environment (Houser 2014).

While these contrasting positions construct telling arguments about authority and authorship, they are also provocations to think about the diverse ways in which images articulate the environment and the contributions of anthropological research to this process. What might drawings (and coasts) look like from anthropology? Can ethnographic research point toward drawings that build forms of care that take destruction, loss, and violence into account? These questions respond to recent calls for attending to the climate crisis as a crisis of storytelling—to forge narratives that dismantle extractive empires, re-world and root people within places, and craft ethical commitments to the environment (Ghosh 2016; Latour and Woolgar 2013). I attempt to experiment with anthropological narratives by combining comics and text, and simultaneously use ethnography as a starting point to rethink the frameworks of technical drawing and find points of intervention through its transformative potential (Pandian 2019). I redraw images, moments from the drawing process, visual encounters, and conversations around drawings. The experiment follows a forked path: first, I show how ideas of objectivity, accuracy, and simultaneity that are central to technoscientific images are constructed through a range of practices. Second, I look at how the developmental and infrastructural visions and timelines that these drawings seek to bring forth become muddled as land, water, humans, and nonhumans mix at the shore.

1948 MODAK MAYER PLAN

1950–57 CITY LIMITS EXPAND

1964 DEVELOPMENT PLAN

1981 PM LETTER TO COASTAL STATES

1992 BB VORA COMMITTEE

1997 SALDANHA COMMITTEE

DEVELOPMENT PLAN—1991--FIRST CRZ POLICY

1995 SLUM REHABILITATION POLICY

1996 BOMBAY BECOMES MUMBAI

2000 SUKHTANKAR COMMITTEE AMENDMENTS FOR S.R.A

2005 FLOOD

2004 TSUNAMI

2005 SWAMINATHAN COMMITTEE SUBMITS RECOMMENDATIONS

2008 CRZ NEW DRAFT RELEASED PUBLIC DEBATES

MITHI RIVER DEVELOPMENT AND PROTECTION AUTHORITY (MRDPA)

DP REVISION STARTS, —2011--- REVISED CRZ
MCGM APPOINTS NOTIFICATION
GROUP SCE

FIELDWORK 2011–12

2015 SAILESH NAYAK COMMITTEE ALLOWS TOURISM AND LAND RECLAMATION

2012–13 EXISTING LAND USE PLANS RELEASED + "PEOPLE'S VISION" DOCUMENT

2014 DEVELOPMENT PLAN 2014–34

2015 UDRI'S "DUMP THIS DP" CAMPAIGN

FIELDWORK 2017, 18, 19

DEVELOPMENT CONTROL AND —— 2018-- REVISED CRZ
PROMOTION REGULATION (DCPR)

-2019 CRZ

A History of Category Errors

Sometime midway in 2012, there was public meeting in a *koliwada* (fishing village) in Mumbai to rally the Koli community against the risks that the 2011 CRZ posed—namely, loss of land, development rights, and livelihood.[6] The meeting, which was attended by prominent members of the community, was called after the police had arrested protestors during a demolition drive in the koliwada. What used to be a line of homes was now reduced to rubble to make space for a high-rise apartment block. The stage was adjacent to these demolished homes and on it were seated prominent leaders of the fisher community, many of whom held important positions in the National Fishworkers Forum—a national confederation that brings together small-scale and traditional fisher unions and associations from the Indian states—and the Maharashtra Machchimar Kruti Samiti, which links the many cooperative fisher societies and fish markets. A community leader took the stage and issued an ominous warning about a "tsunami" of real estate development that would displace fishers to slum rehabilitation schemes—mass housing blocks where residents of informal settlements are relocated.[7] The Koli community, he said, had brought this threat on itself by opening its settlements to outsiders (migrant workers). He then urged the Kolis to remember their identity as Mumbai's *bhumiputra* (sons of the soil) and develop their lands in a manner that benefitted the community.[8]

Such speeches echo the nativist discourses of regional right-wing parties in Maharashtra and elucidate the link between development, modernity, and class and caste politics (Hansen 2001). Koli refers to a caste group that traditionally fishes, and in the state of Maharashtra, Kolis are officially categorized as "Other Backward Castes." There are many different Koli subcastes and communities, some of which are included on lists used for affirmative action programs, though this varies by state (Nair 2021; Peke 2013). As Sheetal Chhabria (2018) writes, narratives of Koli indigeneity have, since the early twentieth century, given the community a measure of protection from development projects that affected other informal settlements, often displacing residents and dismantling industries. Most koliwadas are historic villages that became a part of larger informal settlements, without any clear lines between the old and the more recent constructions. In a way, the difference is a socially constructed one as villages such as koliwadas have long been kept out of the formal development process, and continually worked on by its residents to suit their changing needs. Before it was revised in 2011, the CRZ's environmental restrictions limited the development in koliwadas.

The first coastal policy, the 1991 CRZ policy, curbed construction in informal settlements along the city's edge. The policy revisions in 2011 altered this status quo by opening coastal lands for development. However, as the 2011 CRZ tied these rights to identity—both of communities and of the settlements. Thus, it pitted communities such as Kolis against migrant communities, while also making a distinction between historic fishing villages and informal settlements. These categories are hard to pin down in a city like Mumbai where approximately half the population lives in dense, hybrid, informal settlements that house diverse communities. Historic maps did little to shed light on the problem as older revenue surveys did not map fisher settlements separately, lumping them instead under the category of "village." This meant that most urban villages appeared under the same category in historic surveys, and it was not an easy task (based on the conditions defined by the CRZ) to distinguish and establish their boundaries.[9]

The categories in the 2011 CRZ can be traced back to policy recommendations made after the 2004 Indian Ocean tsunami. When the tsunami hit the Indian subcontinent, a team called the Swaminathan Committee, appointed by the Ministry of Environment, and Forests, and Climate Change (MoEFCC), was reviewing the coastal policy.[10] The committee's report was deeply influenced by the disastrous event. In its recommendations, the committee painted two different pictures of vulnerability. It noted that residents of informal settlements were at risk from natural disasters given the lack of public amenities and the material conditions of their housing. Based on this material vulnerability, the committee recommended that rehabilitation schemes, which were not allowed on coastal land under the previous policy, be permitted in accordance with Mumbai's development rules and regulations (Swaminathan Research Foundation 2005). Though fishers too lived in informal settlements, the committee regarded them as differently vulnerable, pegging their vulnerability to their close relationship with the coast (Venkataramani 2017). The 2011 CRZ stated that traditional fisher communities would have a greater measure of control over the development of their settlement. Consequently, the policy's categories conferred different identity-based rights to communities living in informal settlements. The Koli community stood to gain greater say over development, but only if they could prove their identity and that of their settlements as traditional fishers and fishing villages respectively.

The Swaminathan Committee based its recommendations on the prevalent notion that indigenous communities share a special relationship with the environment and thus have a greater right over it than other communities. Alongside ecological conservation and sustainable development, the 2011 CRZ listed the protection of the lives and livelihoods of traditional fisher communities as one of its main goals. Shalini Randeria (2003) argues that such a distribution of rights on an ecological premise is highly problematic as it romanticizes indigenous communities. An identity-based understanding of rights and community roles jibes well with management paradigms where indigenous communities are framed as wards who have environmental expertise but are not given any say in environmental governance (Bavington 2011).

At the time that the 2011 CRZ was released, there was no survey or register with information on fishing settlements, their location, or their boundaries. Even conducting fresh surveys was tricky business for reasons beyond the blurred boundaries in informal settlements. During a community meeting, a fisher asked whether a village would count as a koliwada if very few residents fished, and would settlements where people had developed their houses still count as historic settlements?

The Koli community's predicament illuminates the primacy of drawings in governing the environment. It stands out as a pithy example of the ways in which contemporary environmental governance works by dividing, locating, and fixing the landscape along principles inherited from colonial modernity. Cartographic images are central to this enterprise because they produce geographic knowledge and define the physical domain for the exercise of power (Edney 1997). They fix categories in space, and the case of the koliwadas are good examples of the unruly afterlives of these categories.

The importance of scientific cartography in modern environmental management can be traced back to colonial governance practices where maps consolidated visual knowledge of the territory. Empire making and geography share an inextricable relationship, as governance and control were established though a perceived knowledge of a territory based on detailed cartographic representations. As Timothy Mitchell (2013, 9) writes of colonial mapping in Egypt, the cartographic instrument served as a "means of recording complex statistical information in a centralized, miniaturized, and visual form. It was to provide not just a diagram of reality, but a mechanism for collecting, storing, and manipulating multiple levels of information." Because it served as a node for bringing together different kinds of information and putting them in spatial relation with one another, surveys allowed the colonial state to divide, categorize, and refigure the territory as an economic domain (Scott 1998).

The map's capacity to act as a political instrument derives from the vantage point that the image offers—of setting the viewer high above, where eyes can observe, control, and defend territory. From this point, it allows the viewer to perform a *coup d'oeil*–to take, in one view, the entirety of the area mapped and to seize it (Daston 2019). The same vantage point that allows the eye to project across a surface also allows it to project into time—for the map to function as a plan. As a plan, the cartographic image is not just about securing territory but functions as visual medium for undertaking the politics of governmentality: it is a means of defining, locating, and making visible populations, and identifying and separating objects of economic interest (Mitchell 2013). This includes classifying the natural world, separating it from the human, and setting mechanisms of management in place by determining the range of possible outcomes or uses for the spaces, things, and people the plan depicts (Chatterjee 2010).

The most important project of the 2011 CRZ was the creation of coastal management plans, as these drawings enacted the new environmental politics of the coast. The drawings determined what constituted the coast, where it began and ended, the development and conservation potential of its various parts, what counted as ecological resource, and who and what belonged in it. In terms of the coastal policy, the CZMPs functioned both as images of scientific rationality and as images of a state invested in balancing sustainability, community rights, and economic growth. However, they are not the only images shaping the coast; the DP's revision too spurred several activist and community-led projects that sought to intervene in the state's planning process.

In the 1990s, the Indian government introduced key shifts in urban governance that decentralized planning practices. Transparency, increased access to information, and urban reform were integral to these initiatives meant to foster partnerships between community organizations and private developers (Coelho et al. 2013; Nijman 2008). While they were supposed to level the playing field and give marginalized communities a voice in the planning process, their effects were largely uneven. Programs introduced under the banner of decentralization became a means of disciplining real estate markets in favor of the urban elite (Benjamin 2010). For instance, Vineet Mukhija writes a detailed history of slum rehabilitation schemes in Mumbai in which he shows how NGOs used participatory planning to organize the residents of informal settlements in their struggle for housing (Mukhija 2003). However, the institutional pluralism and provisions for participatory planning did not automatically result in better housing or better provisions and rights for the urban poor (Sanyal and Mukhija 2001). Asymmetric effects notwithstanding, these measures changed planning and design practice: they created new forms of urban entrepreneurialism and the space for collaborations between architecture schools, planners, NGOs, and community organizations (Appadurai 2001; McFarlane 2012). The collaborative projects that happened with the DP's revision are rooted in this change. However, they cannot be idealized as bottom-up collaborations that resisted market forces; they were as much about participating in market economies, and urban desires to inhabit a global city had just as asymmetric and violent outcomes as masterplans and top-down policies (Doshi 2013a).[11]

Drawing activity intensified in 2011. While these drawings grappled with identity politics and housing rights, they were also about coastal futures and the architecture and ecology of the city's edge. These maps and plans need to be read in relation to other environmental representations such as infographics, visualizations of coastal change, visual documentation, and graphs. Taken together, they form a collective of techniques, practices, and forms of knowledge to recast the city's edge.

RITESH AND I ARE PEERING AT THE SCREEN, LOOKING AT CLIMATE CENTRAL'S INTERACTIVE MAP. ITS TIMELINE PLOTS FLOOD RISK WITH WARMING TEMPERATURES AND RISING WATERS. WE ZOOM INTO MUMBAI, OUR CURRENT LOCATION.

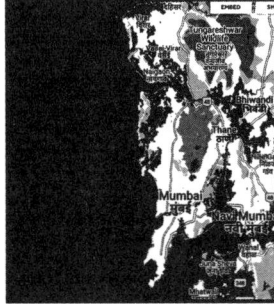

THE MAP ALLOWS US TO ENGAGE IN APOCALYPTIC SPECULATION: WE CAN VISUALIZE WHAT WOULD HAPPEN IF GOVERNMENTS WERE TO MAKE "DEEP AND RAPID CUTS" TO CHECK GLOBAL WARMING. IF WE ARE "LUCKY," SEA-LEVEL RISE IS CONTAINED.

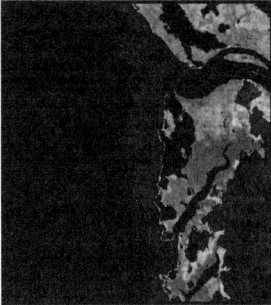

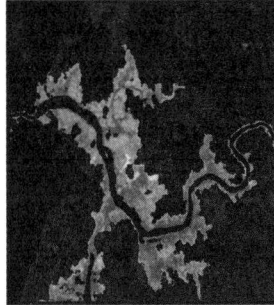

WE MOVE THE CURSOR TOWARD OTHER SCENARIOS: THAT THERE ARE LITTLE TO NO CUTS, AND THE MAP CHANGES AND NOT MUCH IS LEFT OF THE CITY THAT IS NOT UNDERWATER.

RITESH TURNS TO ME AND SAYS, "LOOK, [HE IS POINTING TO MY NEIGHBORHOOD] IS ABOVE WATER. YOU GUYS WILL BE SAFE . . . BUT THE REST OF US ARE GOING TO HAVE TO THINK ABOUT MOVING TO NEW BOMBAY AND OUR HOUSES WILL BE WORTH NOTHING," HE CONCLUDES WITH A LAUGH.

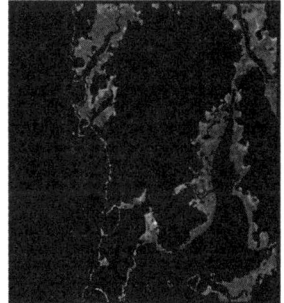

AS LAND DISAPPEARS, OTHER THINGS EMERGE FROM WATERY DEPTHS: NEW TOWN CENTERS, AND NEW REAL ESTATE MARKETS. THEY MINGLE ON THE SCREEN WITH QUESTIONS OF SURVIVAL AND LOSS.

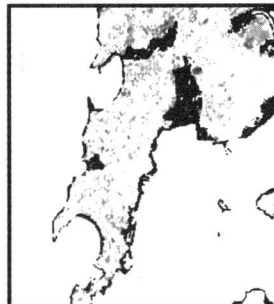

CLIMATE CENTRAL'S MAP IS ONE OF MANY VISUALIZATIONS THAT PROJECTS POSSIBLE FUTURES ON THE PRESENT LANDSCAPE.

IN MUMBAI'S CASE, ITS AQUEOUS FUTURE LOOKS A LOT LIKE ITS ARCHIPELAGIC PAST (ANAND AND TERENS 2019).

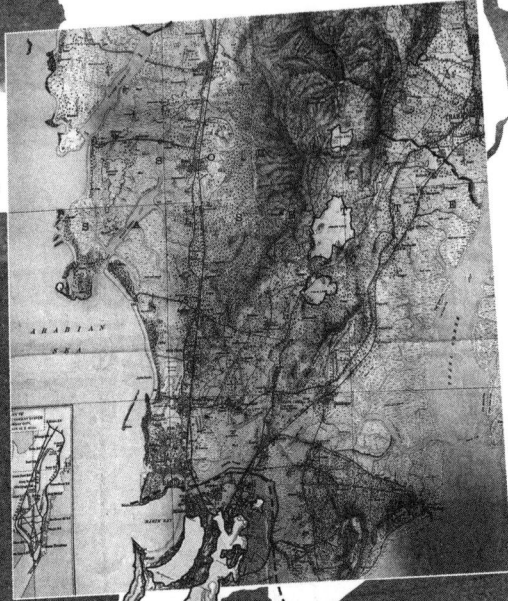

Mumbai Suburban District

Bombay City, Bombay
Suburban, and Thane
Districts. Published
under the direction of
Lt. Col. C. P. Gunter. 1926.

Copy of a map of the
Islands of Bombay and
Colaba prepared for Mr.
Murphy, 1843. In: *Materials
Towards a Statistical
Account of the Town and
Island of Bombay: Vol. 3.*
1894. Government Central
Press, Bombay

Island City

5 KM   10 KM   N

FROM THE LATE EIGHTEENTH CENTURY ONWARD, THIS WATERY LANDSCAPE
WAS FILLED IN OVER SEVERAL RECLAMATION PROJECTS TO PRODUCE THE
PRESENT-DAY METROPOLE (DWIVEDI AND MEHROTRA 1995; PRAKASH 2010).

BACK ON THE SCREEN, WHERE THE RISING SEA LEVEL RECLAIMS THE CITY,
MANY OF THE KOLIWADAS REMAIN. THIS IS NOT SURPRISING AS THEY
EMERGED FROM HISTORIC VILLAGES THAT EXISTED LONG BEFORE
RECLAMATION PROJECTS BEGAN AND HAVE TRANSFORMED WITH THE CITY.
THEIR CONTINUED PRESENCE AS DOTS ON MAPS DESTABILIZE LINEAR
TIMELINES OF DEVELOPMENT, EVEN AS ITS RESIDENTS ACTIVELY SEEK TO
FIND FOOTING IN THOSE PLANS.

FISHING VILLAGES, OR KOLIWADAS AS THEY ARE CALLED, ARE IMPORTANT PARTS OF THIS URBAN FABRIC THAT TUSSLES DAILY WITH LAND AND SEA. THEY ARE CENTRAL TO ROMANTICIZED VERSIONS OF MUMBAI'S HISTORY, WHICH TRAP THEM IN THE PAST. FOR INSTANCE, IN *MIDNIGHT'S CHILDREN*, SALMAN RUSHDIE TELLS AN OFT-HEARD TALE: BEFORE THERE WAS THE CITY, THE FISHER COMMUNITY KNOWN AS THE KOLIS INHABITED THE ARCHIPELAGO THAT BECAME BOMBAY (LATER MUMBAI), AND IN A CHANGING CITY, THE KOLIS, AND KOLIWADAS REMAINED UNCHANGING.

NARRATIVE FOUNDATIONS

Clothes line
Fish-drying area

Terrace tin roof replaced two years ago

Water tank and water supply line added 20 years ago

Second floor added in the late 90s. Occupied by cousin-brother and family

Staircase to the upper story shared with the neighboring unit

Brick wall raised over stone

Water line shared between four households

THIS IS HOW MOST STORIES ABOUT MUMBAI BEGIN: KOLIS ARE DEPICTED AS PEOPLE WHO WITNESS THE BIRTH OF THE CITY AND KOLIWADAS AS SITES THAT HAVE REMAINED ANCHORED IN HISTORY, OUT OF TIME WITH ITS PRESENT. HOWEVER, THE HODGEPODGE STRUCTURES TELL OTHER STORIES. RATHER THAN THINKING ABOUT HOW THEY DESCRIBE AN EXCEPTION TO THE PLANNED URBAN SPACE, I FOLLOW VYJAYANTHI RAO (2007) IN TAKING THEM AS "SELF-PLANNED COMMUNITIES" WHOSE DENSITIES DESCRIBE NOT JUST A CONDITION, BUT SOCIAL RELATIONSHIPS AND PROXIMITIES THAT CHARACTERIZE URBAN EXPERIENCE.

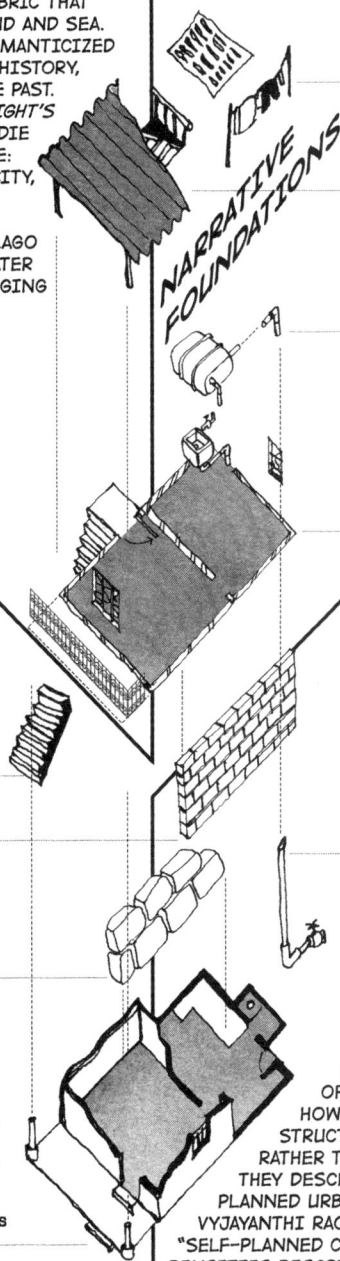

Old stone walls

Baby Koli's house appears in surveys drawn in the 1960s. While its footprint remains the same, it has been remodeled over the years to make room for its residents' needs.

AS MAURA FINKELSTEIN (2018, 954) WRITES OF WORKERS' HOUSING IN MUMBAI, SITES SUCH AS KOLIWADAS "QUEER" TIME BECAUSE THEY RESIST "THE DRIVE TO CONFORM TO THE CONTEMPORARY URBAN TRAJECTORY OF DEVELOPMENT AND DISPLACEMENT, EVEN AS TENANTS STRUGGLE TO REMAIN CONNECTED TO MEANINGFUL FORMS OF HABITATION." THEIR TEMPORAL QUEERNESS IS EVIDENT IN THE VERY BUILDING BLOCKS OF THE SETTLEMENT: IN THE LAYERS OF TIN SHEETS, CAST IRON RAILS, AND BRICK WALLS RAISED OVER OLDER STONE WALLS.

MY FATHER-IN-LAW BUILT THIS HOUSE. OR MAYBE HIS FATHER, I'M NOT REALLY SURE. BUT FOR OUR MARRIAGE, THEY HAD THE WHOLE HOUSE REPLASTERED.

BACK THEN, THE HOUSE WAS JUST THE TWO ROOMS YOU SEE BEHIND ME. THERE WAS A BIG OPEN SPACE IN FRONT OF THE HOUSE - NOT LIKE IT IS RIGHT NOW.

TRACHYTE, AN IGNEOUS ROCK THAT WAS QUARRIED IN SALSETTE (NOW MUMBAI SUBURBS). TRACHYTE IS ALSO CALLED ENGLISH BRICK OR MALAD STONE FOR THE SUBURB WHERE IT WAS FOUND. ALONG WITH BASALT AND RHYOLITE, MALAD STONE WAS SENT SOUTH TO BUILD THE EDIFICES OF THE COLONIAL PORT CITY.

IN THE OPEN SPACE, THEY DUG A BIG PIT WHERE THEY MIXED JAGGERY AND DUNG AND LIME. . . . WITH SOME OTHER THINGS. . . . WHO KNOWS WHAT ELSE THEY PUT?

THE HOUSE HAS CHANGED; WE HAVE THREE FLOORS. BUT THE OLD HOUSE IS STILL HERE, WE JUST BUILT ON TOP OF IT. THE WALLS ARE STILL THE SAME.

ALMOST ALL OF THE OLD HOUSES ARE STILL HERE. THAT IS BECAUSE THEY WERE MADE OF ENGLISH BRICK.

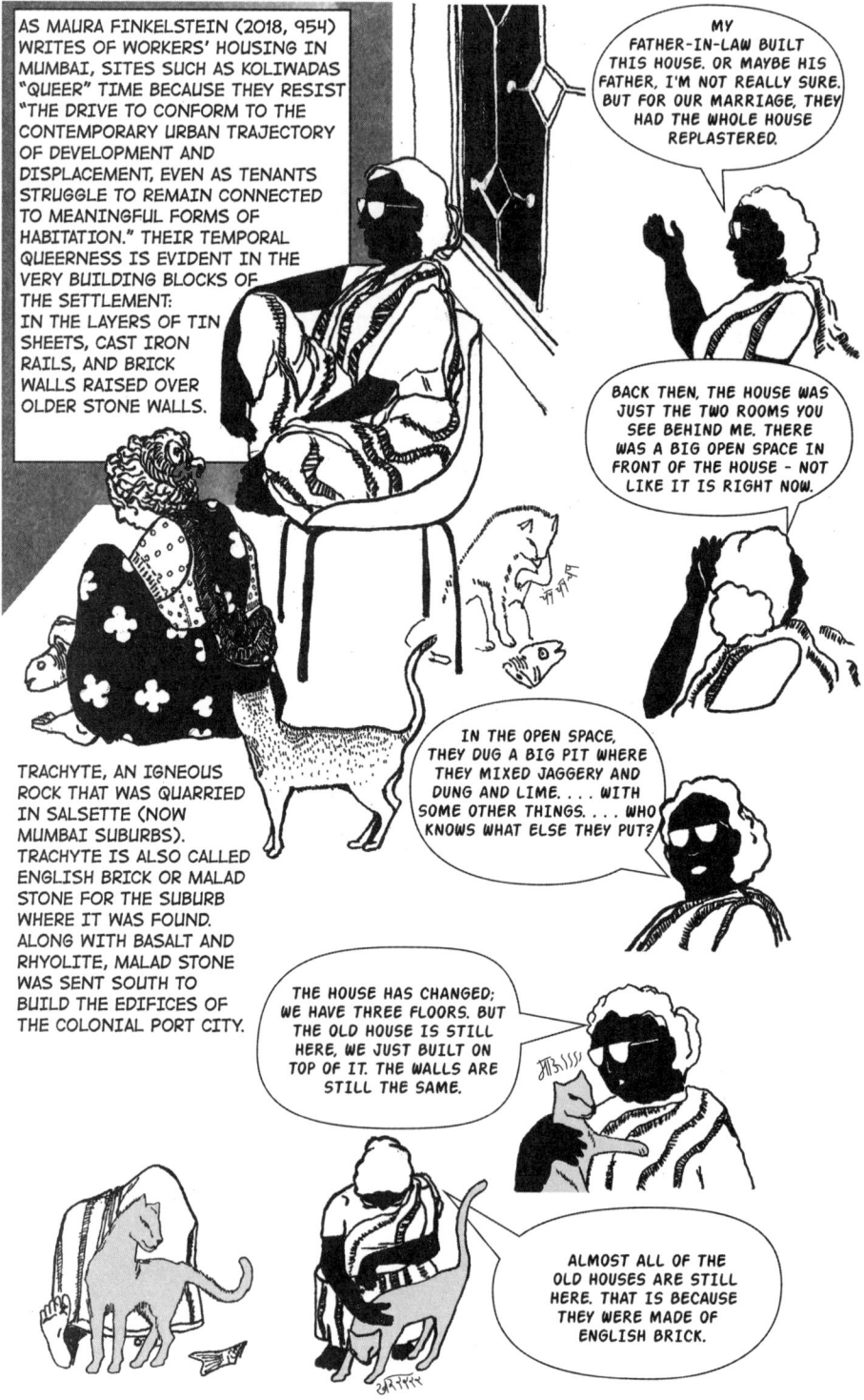

The Technovisual Spectrum of the Climate Crisis

This book is not one that threads a coherent lineage of scientific images and imaging technologies. For instance, while histories of mapping and planning are certainly important subjects to consider, and I do draw on them, I am more interested in the way maps and plans work as material objects, and in relation to other images, technologies, and datasets. Paying attention to these constellations at the shore is a way to consider environmental politics beyond simplistic, oppositional terms such as development versus conservation, or state and non-state. Instead, images can exist along a political spectrum, shifting their positions through their linkages with instruments, infrastructure, institutions, beings, and communities. Thinking in terms of a *technovisual spectrum* moves us away from any simplistic understanding of the image politics of the climate crisis and simultaneously raises complex questions regarding political capacities, authority, and authorship.

I build on social studies of technoscientific images that attend to their distinctiveness and ask what makes them different from other images (Burri and Dumit 2008). These studies delve into their effects both inside and outside communities and spaces of expertise. In the lab, and more broadly speaking, in the context of scientific practices, images are integral to the production of scientific knowledge and networks. They render the material world, objects, and beings as inscribed information, and yet they are not mere recording devices (Latour 1986). For example, Latour and Woolgar (2013) note that the laboratory is not just constituted through its sophisticated instruments or the scientists who work in it, but also through inscriptive practices, such as recording and transmitting empirical observations of creatures or chemicals. These graphic and textual inscriptions allow science to travel, to turn specimens into mobile information and create scientific networks.

As Kathryn Henderson's indispensable work (1998) on technical drawing shows, technical images are not "rigid models" and are able to do many things: they are communicative practices, and they constitute spaces of work and shape disciplines by codifying knowledge. This allows them to function as "boundary objects" (Star 2010)—objects that are established enough to be read by experts and nonexperts, and plastic enough to bend to their various needs. Beyond this interpretive flexibility, technical images like the DP and CZMPs have an architecture that makes them useful for communities that may not agree with one another and have completely different desires and ideas. These objects can organize standards and methods of working—for instance, the scale at which plans are drawn and the colors that are used. At Mumbai's coast, images connect amateur mappers and professionals, coastal communities including fishers, planners, and activists, and they mobilize support for projects that transform the city's edge. They do this by virtue of their properties as technical objects, their ability to claim objective truths, and because of digital infrastructures that make image-making capacities available to the public (Harley 2002). Technical images of the climate crisis proliferate—they are the familiar graphs that depict the dizzying climb of the average global temperature, the threatening progress of the waterline, and the maps of land loss. However, as many others have asked: whose loss do they show? What is the nature of the violence they articulate? What is the outcome of the calls to action they articulate? These are important questions because technical images occupy a special position in climate politics. They are rooted in histories of systematic erasure, where landscapes are emptied of relations between humans, animals, plants, and landscapes in favor of neat, governable orders. Paradoxically, they are also offered up as technologies and media that, when used by indigenous communities and activists, can be remade to challenge those very histories (Marzec 2015). Geographic Information Systems (GIS) and their cartographic images are apt examples of the paradoxical lives of contemporary visual technologies. Soon after Google Earth was released in the public domain in 2005, ecologists were quick to seize on this resource as a way of tracking and visualizing conservation efforts that mapped eco-regions or tracked species while simultaneously making these visualizations available in the public domain. These political contestations are often termed as "counter" cartographic endeavors (Peluso 1995). Counter images are thought to be positioned against dominant visual forces that discipline the landscape to developmental visions. They bring back complexity, they demand the inclusion of elements and communities that are not counted, and they rebel against Cartesian calculus. They are framed as drawings produced at the margins of the state, which exist in oppositional relationship with the center. Often, their political articulations happen through experimentation with form, particularly, through a critique of the technoscientific visual language—for example, by dismantling the bird's eye view of urban plans (Mathur and Cunha 2009).

Counter images are important agents that drive climate politics as they are founded on the knowledge of nature-culture continuities that are ignored in the pursuit of scientific ecological management (Orlove 1991). Even when they are acknowledged, these knowledge worlds are often considered secondary to scientific formulations of environmental relationships (Cruikshank 2007). Indeed, community maps have played important roles in resisting environmental destruction, organizing collectives, negotiating landscape transformations, and producing better cities for at-risk communities (Corburn 2003; Padawangi et al. 2016; Parker 2006). However, cartographic control and counter cartography are only two poles of a wide and varied spectrum of political articulations. Just as indigenous communities and environmental activists do not always occupy counter political visions or automatically assume the position of ecological stewards, images too exist along a range.

The images I found and followed in the field did not occupy comfortable polar positions. Nor did they necessarily take apart orthographic perspectives. Instead, they worked from within conventions and formal languages of technoscientific images used by official institutions. They just as often reproduced the state's claims and visions as they opposed it. They countered, but they also complied, corroborated, hedged, and mimicked.

Far from neutral, digital visualization systems are enmeshed in the politics and economic logics of the corporations that make them publicly available (Helmreich 2011). Thus, rather than thinking of them as neutral objects onto which political claims are mapped, I think of images as articulating a technovisual spectrum through their connections with a range of actors (people, plants, animals, and things), media and media characteristics (digital, paper, high and low resolution), technologies (mapping, counting), institutions, and practices. Taken as a spectrum, images bind the complex circuits from which the politics of crisis emerge.

IN JUNE 2009, THE NATIONAL GALLERY OF MODERN ART IN MUMBAI HELD AN EXHIBITION TITLED *SOAK: MUMBAI IN AN ESTUARY*. THE EXHIBITION'S CREATORS, ANURADHA MATHUR AND DILIP DACUNHA, VISUALIZED MUMBAI DIFFERENTLY – NOT AS A BIT OF LAND RISING ABOVE THE SEA, BUT AS A RAIN-FED SPONGE STEEPED IN IT (MATHUR AND DA CUNHA 2009).

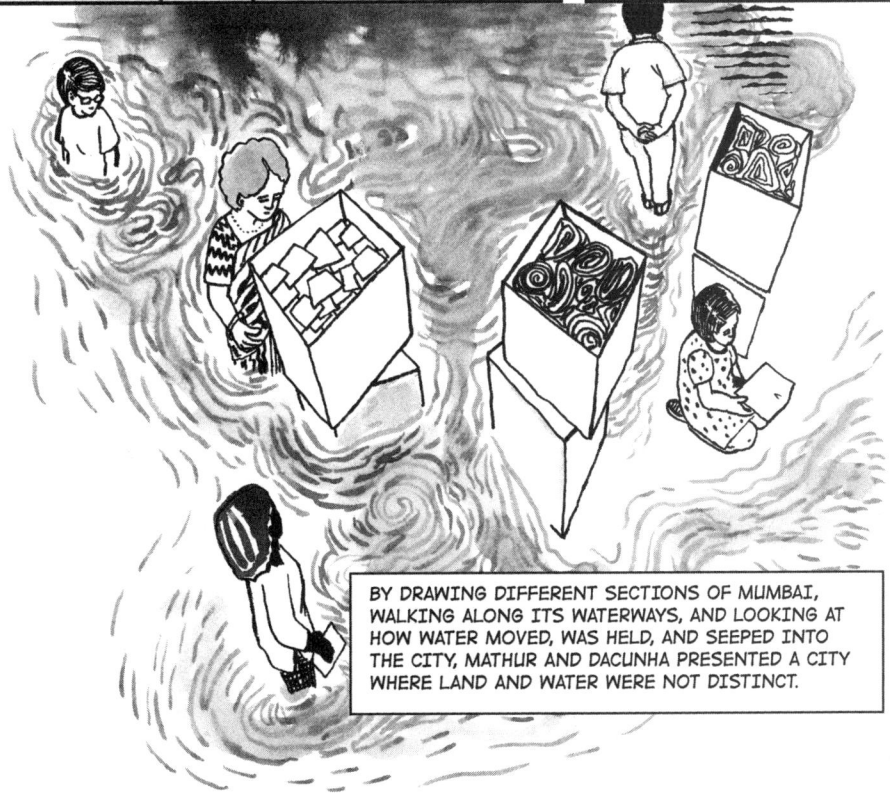

BY DRAWING DIFFERENT SECTIONS OF MUMBAI, WALKING ALONG ITS WATERWAYS, AND LOOKING AT HOW WATER MOVED, WAS HELD, AND SEEPED INTO THE CITY, MATHUR AND DACUNHA PRESENTED A CITY WHERE LAND AND WATER WERE NOT DISTINCT.

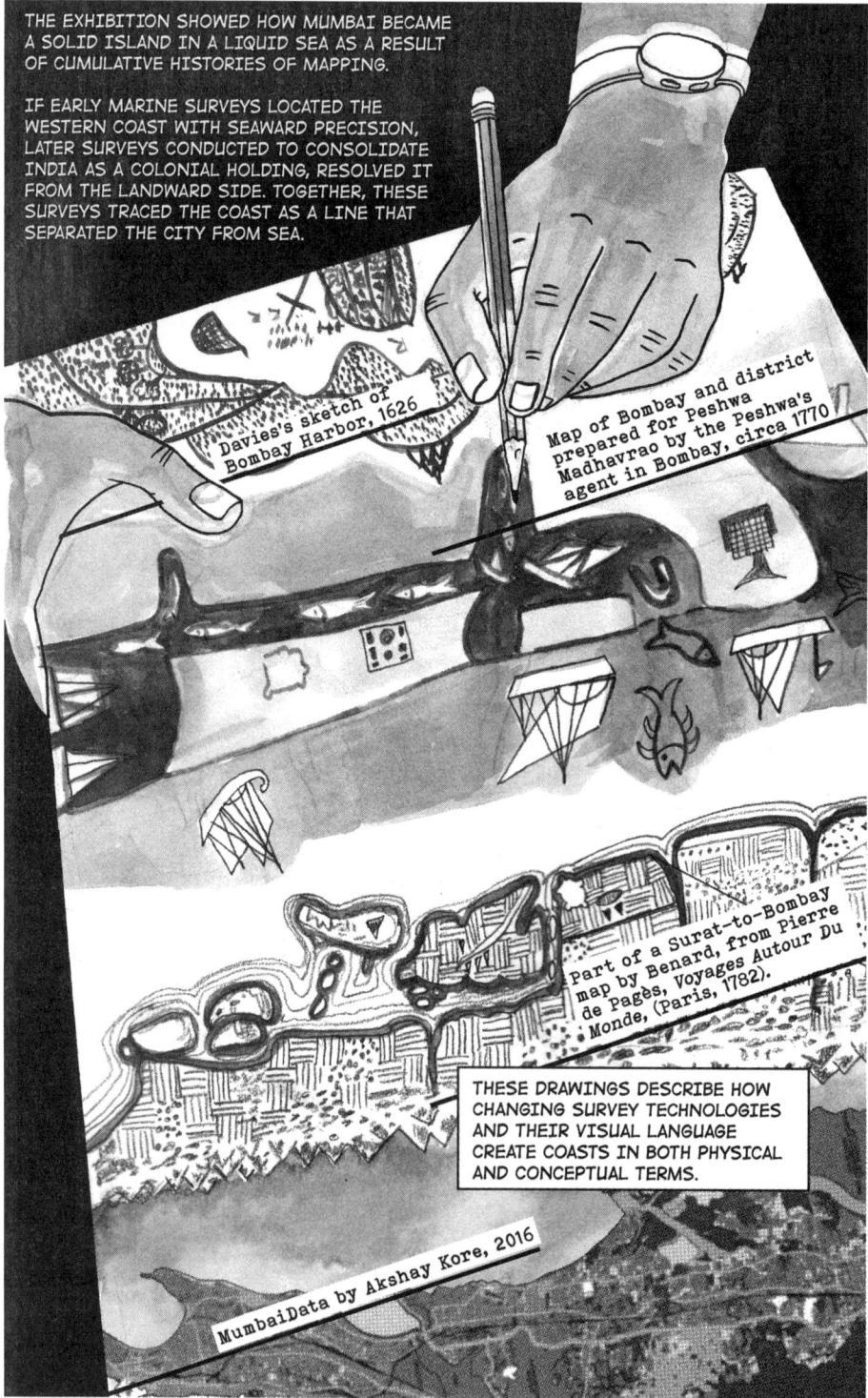

THE EXHIBITION SHOWED HOW MUMBAI BECAME A SOLID ISLAND IN A LIQUID SEA AS A RESULT OF CUMULATIVE HISTORIES OF MAPPING.

IF EARLY MARINE SURVEYS LOCATED THE WESTERN COAST WITH SEAWARD PRECISION, LATER SURVEYS CONDUCTED TO CONSOLIDATE INDIA AS A COLONIAL HOLDING, RESOLVED IT FROM THE LANDWARD SIDE. TOGETHER, THESE SURVEYS TRACED THE COAST AS A LINE THAT SEPARATED THE CITY FROM SEA.

Davies's sketch of Bombay Harbor, 1626

Map of Bombay and district prepared for Peshwa Madhavrao by the Peshwa's agent in Bombay, circa 1770

Part of a Surat-to-Bombay map by Benard, from Pierre de Pagès, Voyages Autour Du Monde, (Paris, 1782).

THESE DRAWINGS DESCRIBE HOW CHANGING SURVEY TECHNOLOGIES AND THEIR VISUAL LANGUAGE CREATE COASTS IN BOTH PHYSICAL AND CONCEPTUAL TERMS.

MumbaiData by Akshay Kore, 2016

Chapters and Methods

The fieldwork for this book happened in two phases—the first was as a graduate student when I spent fifteen months in Mumbai. Most of my work during this time focused on the fishing villages and the politics of planning the coast. When I arrived in May 2011, a couple of fishing villages—Malvani and Moragaon—had reached out to the faculty members at the Kamla Raheja Vidyanidhi Institute for Architecture and Environmental Studies (KRVIA) to see if the school could develop urban plans for these koliwadas as a part of their studio classes. I joined the group of students assigned to study Malvani, a fishing village located in the western suburb of Malad, in the P-North ward of the city.[12] Through the studio, I met Dev Koli, a resident of Malvani Koliwada, who introduced me to his extended family, and this became my primary fieldsite.

As my visits to the village grew more frequent, I began participating in fish-sorting and selling activities at the beach from time to time. Dev Koli and many other fishers practice what is called "nearshore" fishing. They do not venture into deep waters, preferring to stick closer to the shore, and their boats cannot carry more than a few hundred kilos of cargo and catch. In contrast, trawlers and purse-seine netters can carry several tons of catch. In recent times, the government has introduced plans to dramatically increase inland fish farming. The depleted fish stocks and the pressure of competing with trawlers, purse-seine netters, and inland fish farmers have made nearshore fishing unviable for many communities (Bavinck and Johnson 2008).

In Malvani, most fishers did not make their income from fishing alone. Many—especially those who owned land—made money through rent, by working as agents in wholesale markets, and through property development. Nearshore fishing's murky future made the land politics of the CRZ even more important as the fisher community felt it imperative to gain as much development potential and real estate value as possible. Much was at stake in the 2011 CRZ policy—both in terms of development and housing rights and of the ecological future of the coast.

Talking about coastal development in koliwadas is tricky because it brings up questions about ownership and histories of land claims—these are contentious questions to ask of residents in informal settlements. Spending time at the beach and laboring with fish helped establish relationships, which allowed me to talk about these topics. In the beginning, I thought of this fish work as something I did to understand urban politics "out there." However, it was the hours I spent at the beach that allowed me to see the mosaic that reached beyond housing politics and connected the coastal landscape, infrastructure, flora, and fauna.

During my first round of fieldwork, I would often encounter material on the Indian Ocean tsunami and the 2005 flood. I returned to the field for shorter stints between 2017 and 2019. In the second phase, I took an expanded view of the coast and of visual technologies. This allowed me to go back to the material on the 2004 Indian Ocean tsunami and the flood and connect visualizations of weather and the work of marine activists to coastal mapping and planning projects. The book is organized so that it moves along Mumbai's shore, attending to these visual projects that articulate its edge in different ways.

In the first chapter, I follow state-appointed cartographers as they walk along Mumbai's shore to draw the new coastal plan. The 2011 CRZ policy envisioned a straightforward reordering of the coast, implemented through high-resolution satellite maps and coastal surveys that would provide an accurate and objective picture. The chapter provides a detailed account of the survey process and shows how objectivity, accuracy, and immediacy are not inherent to technologies, but are produced through a range of drawing, recording, and communication practices used by surveyors. The shore does not yield easily to the survey or to the instrument's will, and thus, accuracy becomes a visual practice, a quality that surveyors must resolve within the scope of the drawing process. By reconstructing the policies as comic panels, I show how the emphasis on accuracy and image resolution ends up creating a *terranean coast,* which is organized as a set of classified, geo-located elements carefully recorded on the plan. The resulting plan reterritorializes the coast in terms drawn from land governance and renders it as a fixed, terranean entity separated from the sea.

The second chapter explores the ways in which the coastal policy draws in communities that fish in Mumbai's shallow, nearshore waters and the nonhumans that inhabit the city's shore. Nearshore fishing works through a web of shifting relationships and knowledge flows against which day-to-day decisions are made. Economic and kin relations are important but not the sole components of this web, which includes marine creatures, beaches, and fishing infrastructure, the creek, and the sea. These intimacies are changed, and new uncertainties emerge, because of the coastal policy and its attendant classificatory orders.

The drawings in this chapter trace the divergent ways in which small-scale fishing and the CRZ conceive the coast. The CRZ reconfigures fishing communities as stakeholders and as vulnerable subjects who rely on the coast for a living, while the fish are reconfigured as catch, bycatch, and economic or ecological resource. These reconfigurations introduce pressures into the already dynamic field in which fishers operate and fish spawn. However, this coastal reordering does not happen through a straightforward process: to naturalize and legitimize these regimes of coastal management, the state must first determine the boundaries of the coast through coastal surveys. In this drawing process, berms, dunes, mangroves, scrub, and debris—and most importantly, salt or salinity—are signifiers that demarcate new territories and reorganize the liminal space where land and sea meet. These material, vegetal, and landscape elements often resist interpretation through the unexpected ways in which they behave or appear, making it difficult to delineate new boundaries and zones. The chapter follows cartographers as they map the tidelines to fix the coast, papering over the ambiguities caused by nonhuman agency.

The third chapter turns attention toward how data, when visualized as infographics, narrativize and shape the experience of extreme weather events. They rescale the scope of infrastructural systems and, as the extreme becomes the norm, result in a landscape shaped by anomalies. The chapter builds this argument by tracing the memory of the 2005 Maharashtra flood, which, combined with the rising incidence of heavy rainfall, led to a widespread interest in monsoon-related data. The chapter links infographics—different bits of data that are combined to create a visual narrative—with engineered landscapes by following the plan to upgrade Mumbai's drainage systems to handle extreme rainfall. To handle the rising precipitation volumes, the city's complex coastal estuarine systems were reimagined as defined riverine channels, further hardening the edge between the city and the sea.

I redraw these infographics, the stories they tell, and visualize their data flows. While infographics of extreme rainfall give rise to new calculi of infrastructural capacities, other statistical figures chart different sensibilities and urban relations. The chapter delves into amateur meteorologist collectives that gather weather data from a range of sources such as personal weather stations, automated weather monitors and rain gauges, and governmental reports to create visualizations that circulate via social media. The graphics created by the amateur meteorological collectives create a weather-minded public that uses the visualizations to make everyday decisions regarding their movement. The infographics created by the collectives are also important because they build an understanding of the monsoon as a complex phenomenon that works across different scales of localized microclimatic events and the movement and progression of the monsoon across regions.

The fourth chapter attends to the different ways in which future-oriented drawings, such as urban plans, unfold in time as they go through various stages of their creation, drafts, publication, legalization, and circulation. It introduces the second state-led drawing project, the revision of Mumbai's DP, and the unofficial planning initiatives it spawned, which became an important part of the fisher community's struggle. Each stage of the drawing process creates different kinds of openings for political action and articulation, which, instead of grappling with uncertain ecological futures, remain trapped in the attenuated timelines of the urban plan and tied up with questions of identity and belonging that characterize regional right-wing politics. Even as the drawing process offers multiple entry points for political intervention, the cartographic plan's development-oriented focus closes the doors on collective engagement and on the problems of climate change and environmental degradation that exceed planning horizons.

The chapter follows the unofficial plans created by the fisher community in alliance with planners and activists. These unofficial drawings were not about challenging developmental vision—rather, they curbed short-term threats while leaving the room for future political maneuverings. The comics in this chapter play with images central to the Koli community's political articulations. To create new opportunities for action, the fisher community deployed a range of images and image-making practices—particularly, traditions of colonial photography and practices of figuring indigenous bodies. These images allowed them to establish belonging, and when coupled with survey images, they secured them in place. The rich social life of survey images expands the idea of expertise and shows how surveying is not just an encounter between a mapper and the landscape, but an open moment with the potential to shape community and coastal futures.

The fifth chapter continues the conversation on images and time. Using the timelines of garbage and chemical action of matter in Mumbai's dumping grounds, it reconceives the coast as an amalgam of nonhuman activity and interactions. Built on the estuarine coast, these dumping grounds are supposed to follow a fantastical trajectory where they are harnessed for energy and converted to saleable land. Within the decadal timelines of these waste management programs, the complex chemical mix that is waste is separated and undergoes a controlled transformation after which it is classified as inert and safe to render into landscapes of human design. However, waste matter continues to work, building continuities that attest to the ways in which waste is now a fundamental aspect of coastal chemistry. When seen as a coastal substance, waste presses against the idea of the coast as a space made of discrete entities. Waste and waste's encounters with fish, fishers, scientists, and flora and fauna directs our attention to how plastic, chemical pollutants, and sewage are not just happening to the coast; they have long animated Mumbai's coastal atmosphere, water, soil, people, and creatures.

The chapter follows marine life advocates who lead public walks along Mumbai's shore to reveal how plants and animals persist despite/because of the changing coastal chemistry. They meticulously document marine life through photographs, encouraging a citizen-science movement of learning to recognize and see signs of life along a shore that is otherwise characterized as a dead zone. The online photographic archives and the visual encounters these enable craft a possible posthuman politics. Instead of thinking of the creatures as occupying space designed for human needs, the archive recasts the coast as an outcome of waste's transformations and interactions. In the process, the images suspend the viewer's anthropocentric view; they create the possibility for thinking about loss beyond the language of complete destruction, and for thinking about coastal futures as futures crafted by the actions of waste.

Drawing Predicaments

Much of the anthropological work we encounter on page is text. This is changing in recent times, with growing interest in comics as a medium for conducting and expressing ethnographic research. In a series of essays published on *Cultural Anthropology*'s website, Dimitrios Theodossopoulos (2022) writes that this emergent "graphic-ethnographic-practice" moves away from thinking about drawing as a tool for expressing ethnographic analysis. Instead, it takes graphics as grounds for new opportunities for collaboration, analysis and reflection, and public dissemination.

The form that this book takes and the image-text relations in it are, in some part, an outcome of such desires. However, this does not entirely answer the question that is often posed to nontextual (or perhaps more-than-text is a better term here) anthropology: why must a thing be drawn (or filmed or sung or danced)? In this case, the desire to explore new mediums is a misleading answer because drawing is decidedly not novel. As Lynda Barry (2019, 1) observes, the "ability to write could only come from our willingness and inclination to draw." When this question is posed, what it also conveys—without the words—is that there is a willingness to dive into text—to forget its channels of abstraction, and to take on icons that forge sound, words, and meaning. Pictures, as Scott McCloud (1994) draws, make meaning differently from texts. To fully engage more-than-textual scholarship of any kind also requires the readers to embrace unfamiliar channels of abstraction. In this case, it requires embracing the dual nature of comics, its circuits of meaning making and storytelling, and the willingness to see what is shown, rather than being told what is pictured. Thus, instead of dwelling on how comics might separate this work from others, I approach these questions not as ones that are posed of comics, but as questions that might be good to ask of any work: why does a thing take the form it takes and what is that form doing?

In their book on the work of sound in audio-visual experiences, Michel Chion (2019) writes that cinema is often mistaken as a visual experience (i.e., one goes to "see" a movie). In the process of giving importance to the visual, sound, which is vital to the cinematic experience, is rendered as secondary. Chion suggests turning off the screen and listening only to the sound to understand how it makes cinematic narrative and its relation to the images on screen. Like cinema, comics too are mediums that operate in two channels: image and text. They tell stories by balancing and constructing complex relationships between pictures and words (McCloud 1994) and much happens by showing rather than telling. This is an important aspect of graphic ethnographies: for example, in *King of Bangkok* (Sopranzetti et al. 2021), we see how a blind man senses the city through sound. We read this in the play between words that make up sounds and planes of color that translate that as a blind man's perception of that sound. None of this is written and explained. In *Lissa* (Hamdy and Nye 2017), the creators do not tell us about the complex ways in which illness reshapes the experience of time, they show it through the narrative drawn on page. And in *Forecasts* (Schuster 2023) the speculative futures and anxieties of microfinance play out in the structure of the panels and pages and in the motifs that carry through the book. In making the reader follow along without being told to follow along, these picture-word stories articulate thought, happenings, narrative, experience, and concepts in ways that are not new, but they are different from text (Sousanis 2015).

During my time in the field, I would draw sketches of the fish, the fishing village, what I saw at the beach as I waited for the boats to come back, doodles of people, drawings of houses in the village and their histories, and diagrams of institutions and institutional processes. These drawings were catalysts for thought that translated into text (Douglas-Jones 2021; Taussig 2011), but they also did more—they produced more drawings. Just like technical images, these too are rooted in Mumbai's visual politics and structure the comics in important ways.

The CRZ's and the DP's revision created much uncertainty, because of which, conversations on land claims or coastal change were not always welcome. Moreover, the government's policy on geospatial images at the time classified maps of the coast as sensitive documents. This meant that the surveyors and engineers I spoke to could not share these images. My participation in most of these projects and access to the information was predicated on the promise that I would not ask for any maps or make copies of them. I was also requested not to discuss "CRZ politics"—this was a request made by the official who had allowed me to observe the process of creating the CZMPs as they themselves were under a lot of scrutiny at the time. Thus, I felt that I could not construct a narrative structured around protagonists or talk to the surveyors or municipal officials about their takes on the policy and its outcomes. Instead, I have focused on the technologies and drawing processes. This meant that many of my comics are structured around redrawing the technical images or information I worked with in the field, without including any information on the actual location or the identities of the surveyors. While the drawings are my own, I often used diagrams to understand unfamiliar concepts and processes—some of my interlocutors would use sketches to explain things to me, and those small collaborative acts structure the pictures in this book in important ways.

The amphibiousness of comics (Sousanis 2015) is an aspect that is not just fundamental to analysis or its communication, but also to the subject of this book. The extractive landscapes we inhabit are a product of anthropocentric drawings that plan futures in terms of human timescales. Dismantling these extractive regimes requires dismantling the visual narratives they are founded on. In other words, the regimes of climate change are, at least in part, outcomes of visual work. This book builds on work by scholars who draw attention to the crisis of imagination that keeps alive the idea that climate change can be addressed through solution-oriented, short-term future making. In the case of technical drawings this means imagining different lives, forms, and representational frameworks, and drawing them differently.

While they may appear as commonplace tools for technocratic governance, technical instruments and objects are also "wonder-inducing" objects that can awe, dismay, confuse, and "open up the familiarity of what seems straightforward" (Ballestero 2019, 32). Thus, each chapter attends to different possibilities that drawings offer and what they do. In the beginning, the book shows how the coastal policy, and its attendant representational practices, systematically reorganize the coast. Simultaneously, it positions small-scale fishing as a conceptual counter to this technocratic vision. It sows moments of ethnographic doubt in drawings that are rooted in ideas of objectivity and accuracy. In renarrating infographics, it explains how coastal geographies materialize through data stories. In thinking with chemical timelines of plastic, it positions waste as an agent that confronts us with the undeniable presence of harmful things that cannot simply be managed or even buried away. It tries to build a case for drawing more, drawing from anthropology, and drawing differently—in ways that reckon with the long arc of the climate crisis lapping against the shore.

TERRANEAN COASTS

Mapping, Accuracy, and Drawing
the Landward Edge

ACCORDING TO AN ATLAS ON SHORELINE CHANGE PUBLISHED BY THE INDIAN SPACE RESEARCH ORGANIZATION AND THE DEPARTMENT OF WATER RESOURCES, THE INDIAN COASTLINE IS 7,516 KM LONG. OF THIS, 5,422 KM RUNS ALONG THE MAINLAND, WHILE THE ISLAND TERRITORIES ADD ANOTHER 2,094 KM.

IN OTHER DOCUMENTS, THE COASTAL LENGTH VARIES. FOR EXAMPLE, IN A PILOT REPORT PRODUCED BY THE WORLD RESOURCES INSTITUTE, THE INDIAN COASTLINE IS LISTED AS 17,181 KM LONG.

THE LARGE DISCREPANCY IN THESE MEASURES IS THE RESULT OF A PHENOMENON CALLED THE "COASTLINE PARADOX," WHICH EXPLAINS THE RELATIONSHIP BETWEEN COASTAL LENGTHS AND MAP SCALES.

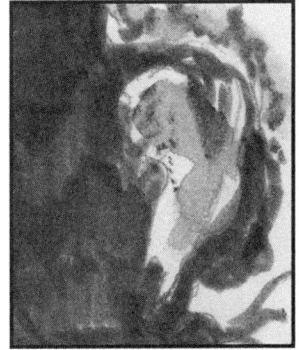

THE PARADOX IS THE OUTCOME OF THE IRREGULARITIES INHERENT TO THE EDGES OF LANDMASSES AND THEIR FRACTAL-LIKE NATURE.

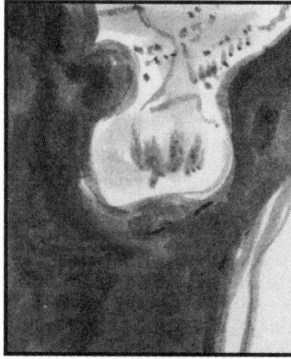

LIKE FRACTALS, AS ONE ZOOMS FURTHER AND FURTHER INTO THE COAST, NEW FOLDS AND BENDS APPEAR THAT KEEP ADDING TO ITS LENGTH.

THUS, THE LENGTH OF ANY COAST DEPENDS ON THE SCALE OF THE RULER USED TO MEASURE IT: THE LARGER THE RULER, THE LESSER THE DETAIL, THE SHORTER THE FINAL LENGTH.

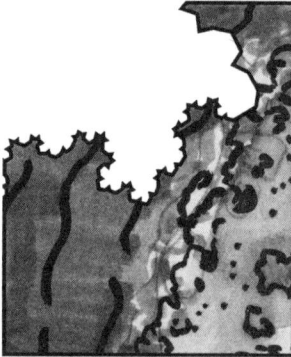

THOUGH THE GOVERNMENT'S ATLAS DOES NOT SPECIFY THE SCALE THAT IT USED TO MEASURE INDIA'S COASTLINE, IT IS CLEARLY SMALLER THAN THE 1:250,000 SCALE USED IN THE PILOT REPORT.

THE COASTLINE PARADOX DEMONSTRATES THE IMPOSSIBILITY OF MAPPING OR PROVIDING AN ACCURATE MEASURE OF THE COAST.

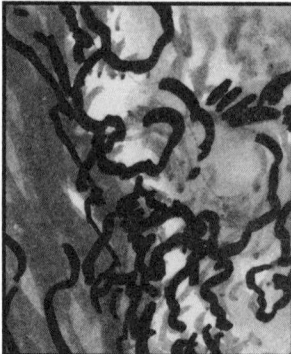

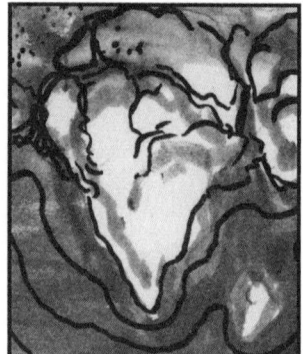

In 2011, the Indian government released a new version of its Coastal Regulatory Zone (CRZ) policy, which governs India's entire coastline.[1] The policy divides the coast into different zones based on population density, infrastructure, and ecological sensitivity, and this classification determines their conservation status or development potential. Though the 2011 CRZ followed the schematic structure of its predecessor, the 1991 CRZ, it added many new rules and subcategories within the zones, which were tagged to specific areas, communities, locations, and environmental features.[2] To implement the policy, the government had to first identify those zones, communities, and features in the landscape. This responsibility fell to the state governments, which contracted agencies to survey the coast using high-resolution satellite imagery and global positioning system (GPS) technologies. Based on these detailed surveys, the state governments began preparing Coastal Zone Management Plans (CZMPs)—cartographic drawings that locate, classify, and draw new boundaries to subdivide the coast. CZMPs decide how the newly classified territorial units can develop—where things like infrastructure can be built or areas that can be appropriated for "public benefit."

Coastal plans and the survey technologies used for making them mark a significant shift in the geographic and conceptual understanding of the coast. Because the CZMPs focus on resource development and property rights, they end up taking a *terranean view* of the coast. These drawings are premised on a fundamental divide between land and sea, ignoring the ways in which the coast is a liminal site where elements mix (McCay 1947). Drawn on a cartographic canvas, the images solidify and anchor the coast in place, rendering it as a fixed spatial entity seen largely from the landward side, a view that has had drastic ecological consequences (Menon et al. 2015; Menon and Kohli 2008). The outcomes of these plans are particularly dire for coastal communities as the technocentric modernizing projects they usher change the physical, economic, and ecological makeup of the shore (Dewan 2021). While I attend to nearshore worlds in the next chapter, this chapter steps back to take in the coastal canvas of the cartographic plan. I look at the ways in which ideas about technological accuracy and scientific objectivity play out in the process of making these drawings and the coasts that take form as a result.

Like other policy initiatives in recent times, the 2011 CRZ was pitched as a bureaucratic apparatus that would be readily accessible, transparent, and accurate through its use of media technologies (Mazzarella 2006). The CRZ's charisma lay in the cartographic plan—particularly, a plan made at a high enough resolution to depict the boundary lines of individual properties. The highly accurate CZMP was supposed to render the coast self-evident in a manner that would allow residents to immediately grasp the new development restrictions or possibilities for their property and their locality. The idea of a self-evident coastal order is an outcome of advances in digital mapping technologies and their wide social reach. Accuracy, affect, and recognition—aspects through which reterritorialization happens, are linked to features such as scale and resolution that structure the visual language of cartographic images. However, accuracy and its relation to belonging, objectivity, or even the immediacy of recognition are complex.[3] They are linked to a long history of reterritorial- izing and governing through cartography—a history that holds particular significance in the region.[4]

The force a cartographic image gains because of its capacity to present an objective picture is, as is the case with other scientific images, the outcome of a combination of elements and practices. Objectivity, as Daston and Galison (2010) write, is produced by training scientific practitioners in representational traditions that allow them to recognize use-specific elements and conventions, and it is also produced by shaping sight in the seeing subject who can discern the knowledge transmitted via the image as "objective." In the case of maps, this objective force is coupled with "accuracy"—the idea that the map can adequately stand in for the area it depicts such that a viewer can gain a spatial understanding of it without ever visiting the location (Thongchai 1994).

In the 2011 CRZ notification, the government placed much confidence in satellite technology. The CRZ remade the coast by overlaying a new classifi- catory order on it, and the 2011 CRZ assumed that the availability of high- resolution mapping and imaging technologies would make this coastal reinvention a relatively uncomplicated process. However, the path from satel- lite imagery to a digital plan was hardly as straightforward as imagined in the policy. Given that the CZMPs are essential to the CRZ—indeed, as authors of a review report published in 2006 remarked, it is not possible to implement the CRZ without these plans (Mukherjee et al. 2010)—it is important to unpack exactly how accuracy manifests in the drawings, how it shapes the public life the CZMPs, and the kind of coast that is produced as a result of this preoccupation with accuracy.[5]

As I followed the surveyors mapping the coast, it was clear that accuracy was much less the inherent capacity of survey instruments than an elabo- rate fabrication; it was produced in different ways in the process of creating the CZMPs and through negotiated encounters between the surveyors and what they found on ground. Despite the contradictions that riddle the idea of accuracy and its various manifestations, it was also the keystone that locked together policy, image, and landscape. The CZMPs conceptual emphasis on accuracy reinforced the divide between land and sea and brought a terranean coast into view.

SATELLITE LAMPS IS AN ARTWORK CREATED IN 2015 BY A TRIO OF DESIGN RESEARCHERS, EINAR SNEVE MARTINUSSEN, JØRN KNUTSEN, AND TIMO ARNALL, AT THE OSLO SCHOOL OF ARCHITECTURE AND DESIGN.

THE WORK COMPRISES A SET OF LAMPS FITTED WITH GPS RECEIVERS, WHICH COMMUNICATE WITH SATELLITES PASSING OVERHEAD.

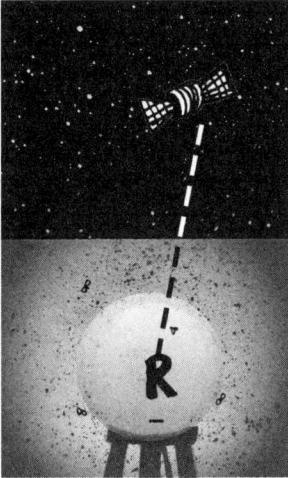

DEPENDING ON THE FLUCTUATIONS IN THE SIGNAL, THE LAMP LIGHTS WAX AND WANE . . .

. . . SOMETIMES THEY FLICKER OR DIE, AS THE SIGNAL CUTS OUT ALTOGETHER.

THE INCONSTANT GLOW OF THE LAMPS ILLUMINATES THE ASSUMPTIONS ABOUT ACCURACY AND IMMEDIACY THAT SHAPE GPS AS A TECHNOLOGICAL AND CULTURAL OBJECT.

GPS POSITIONS ARE TRIANGULATED THROUGH SIGNALS FROM THOUSANDS OF SATELLITES ORBITING OVERHEAD. THESE SIGNALS HAVE UNEVEN STRENGTH AND QUALITY.

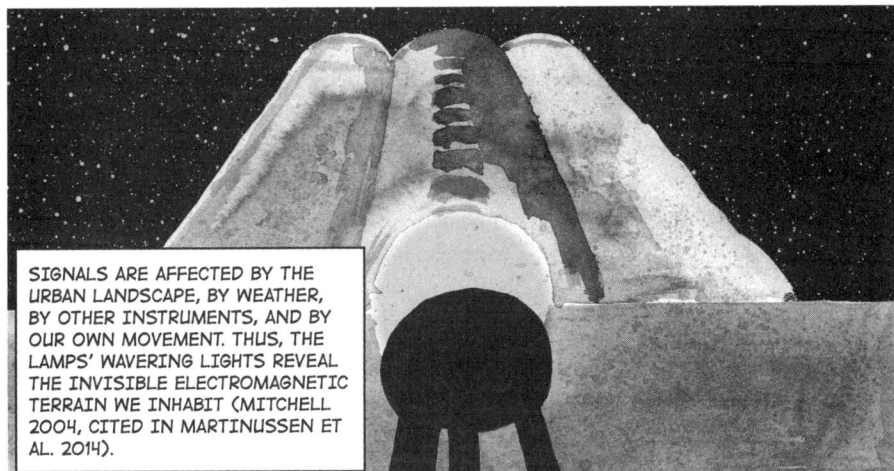

SIGNALS ARE AFFECTED BY THE URBAN LANDSCAPE, BY WEATHER, BY OTHER INSTRUMENTS, AND BY OUR OWN MOVEMENT. THUS, THE LAMPS' WAVERING LIGHTS REVEAL THE INVISIBLE ELECTROMAGNETIC TERRAIN WE INHABIT (MITCHELL 2004, CITED IN MARTINUSSEN ET AL. 2014).

THE CONFIDENT DOT ON THE MAP IS SHOT THROUGH WITH UNCERTAINTY, WHICH WORKS ITS WAY INTO RELATIONSHIPS THAT BIND HUMANS, TECHNICAL DEVICES, AND THE NETWORKED CITY.

Drawing Accuracy

For two weeks in early 2012, I was walking up and down creeks, beaches, and channels in Mumbai with the staff of the Institute of Remote Sensing (IRS) at Anna University in Chennai. Anna University's IRS was commissioned by the MCGM to create the plans for Mumbai's coast, as it is one of the few institutions approved by the MoEFCC to prepare CZMPs. The team working on the project had divided themselves into smaller groups, each of which consisted of a research member from the university, one engineer from the MCGM, a temporary worker who had been employed to assist in taking the readings, and last, a driver. The engineer, Mr. Sudhir, who had arranged for me to come along for the survey, told me that given the time crunch, the presence of municipal officials would give us access to areas that we would otherwise have no hope of entering without prior permission.[6]

The permission to follow the coastal survey came with some strings attached. According to the central government's map policy, coastal areas were classified as "sensitive" zones. It was not possible to freely circulate or publish high-resolution maps without obtaining clearance from the Survey of India (SoI). I was allowed to participate with certain caveats: I was to not copy any information, take pictures without permission, or talk about "CRZ politics" as this was a contentious subject given the land claims in Mumbai. It was important that I adhered to these stipulations as the MCGM official who had allowed me to observe the survey had obtained clearance on my behalf. This should not be taken to mean that the municipal officials or the survey officers believed survey technology to be apolitical; this was far from the case. Their reluctance stems from their acute awareness of the political contentiousness of survey technologies.

On the night before the first survey, I received a text message that asked me to arrive at a government hospital in the western suburbs. In the morning, one of the professors in charge of the survey and a staff member explained their survey method as we sat among a crowd of people waiting to enter the emergency room. One of the team leads said that they would obtain highly accurate locational data using the differential geospatial positioning system (DGPS). In this system, the surveyors install an instrument, called a main or reference station, at a fixed location. Here, it continuously receives its location from satellites orbiting overhead, allowing the instrument to calculate the average error in the satellite data. When the surveyors go out in the field to collect readings, these fixed stations help raise accuracy by correcting the errors in the satellite data (caused by weak or unsteady signals).

DGPS does help raise survey accuracy. However, while walking with this team of researcher, engineers, and surveyors, it became clear that GPS accuracy was just one part of the picture. Making the CZMP was not just a matter of taking geographic readings and transferring them on a visual interface—it involved different technologies and devices, reference maps and surveys, recording systems, and constant movement between the landscape and the recorded data. Thus, the resulting image, the CZMP, was not a singular object, but the outcome of a collection of apparatuses and elements linked together in a long chain of production (Latour 1999; November et al. 2010) of which the surveyors and municipal officers were important parts.

The team had two tasks to complete in the field: the first was to map anchor points, which are points where the location is calculated precisely, and this helps fix the drawing's coordinates in a digital geographic information system (GIS). The second was to map the hightide line (HTL) along the city's coast. After gathering this data, the team would then combine them with existing cadastral or topographic surveys and other databases, such as tide information.[7] To do this, researchers had divided the city into a broad grid. At the center of each grid was a main station, and the dimensions of the grid were based on the transmission range of these devices. The main stations acted as both transmitters and receivers: they received signals from the GPS satellite networks that enabled them to determine their position with a high degree of accuracy. They also transmitted signals to other devices (held in the lands of the surveyors in the field).

On the first day, we were standing in the shaded parking lot of the hospital and once everyone had assembled, we slowly walked up to the roof, where the main station had already been installed. One of the surveyors explained that the station needed to be in a place that would be open, with nothing blocking the signals it would send or receive. The team then went through an instrument check and walked back down to the parking lot. The surveyors spread out a printed high-resolution satellite image on which they had marked the main station. The group was then split up into teams that would map marked-out areas. I was to join Anbu, a graduate student who was a part of the project; Mr. Gomes, an MCGM engineer attached to the area we would map; and Mr. Jatin, a person hired as an assistant.

Anbu's task that day was to get the location of ground control points, which are important for ensuring that the base image used to draw the map is precisely aligned (geo-referenced) in the GIS software. The surveyors had high-resolution satellite images, which would be used as the base maps. However, the images by themselves had no spatial location, and this is where the control points came in. The ground control points would help ensure that latitude and longitude of any given point on the base image aligned with the GPS coordinate system—that is, they would help align the base image in space. To get these control points, surveyors had to get the precise coordinates of locations that were highly visible on the satellite map. Once the satellite image was imported into the GIS software, the surveyors would then look for those locations, to which they would assign those coordinates.

Once the teams were assigned, the first task was to identify potential control points on the satellite image that were distinct and accessible. The project leader unfurled a high-resolution map of the area assigned to us and motioned Anbu to come over. He quickly marked some areas that he thought would act as good control points. He then consulted Mr. Sudhir who looked them over and eliminated a few that would be hard to access. Once the project lead had left, Anbu pored over the map and along with the engineer, tried to figure out the most efficient routes to the locations. He said that though not all these locations might work out, they were good picks because of the way they stood out on the satellite image. This meant that they would be easy to make out on the ground and easy to spot later once the image was uploaded into the software. Later, in the back of the cab, Anbu was describing how surveyors make choices: they are not just looking for places that they can get to easily, they also want to ensure that those points were legible on the satellite image.[8]

As Anbu began plotting survey points in the field, he would constantly look for locations that stood out in the chaotic landscape and in the image. For example, on the first day he spotted the entrance to a series of tennis courts and went up to the watchman at the gate. After about fifteen minutes we were ushered into an area where several tennis courts were situated behind high chain link walls. Anbu walked into the tennis court nearest to the gate and then positioned the instrument on one corner of the court. At the point where two white lines of the tennis court intersected, he placed the receiver. The receiver had to be held for several minutes in order that it get readings from the satellites orbiting overhead and from the main station. Multiple readings meant that errors could be averaged out. Anbu would also reduce error by picking visible control points. As we waited for the receiver to pick up signals, he motioned me to come over and hold the instrument as he marked the point on the printed satellite image. "See this" he said, as he held the image and pointed his finger at the place where the court appeared in the satellite map. Then he neatly circled the corner where the receiver was now planted. "Now I will have no problem identifying this point later," he said.

On another day, as we were taking the reading at a municipal garden where the jogging track turned at perfect right angles, Anbu fiddled with the instrument and handed the antenna over to the assistant instructing him to stand very still to register the reading. Then gazing into the distance, he said, "When I go back and start working on the computer, I will remember this place and where I was standing when I took this point. Sometimes it just comes back, and you know exactly where the reading was taken. Of course, you also record it because you cannot carry everything in your head and there is no guarantee who will [later] work on which part of the map . . . [that's why] when I am taking a reading during the survey, I have to think about what that image will be, or what that point will be on the screen [when viewed by another person]."

Anbu had to think about correctly marking a point in the reference image and the ground while also considering how that point would appear once he began working on the map, and that is why he would pick the most discernable features in the landscape, such as the corner of a jogging track. Over the next three days, the selection of points followed a similar logical positional and visual pattern: not only were they points within the transmission radius of the station, but also points that stood out on the ground and, correspondingly, on the image, stripping away the need for interpretation as much as possible. Corners were better than curves, high contrast was better than low, and corners of unique features were preferred to those of repeating elements. Every selection was an exercise in reducing as much visual ambiguity as possible.

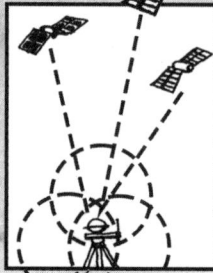

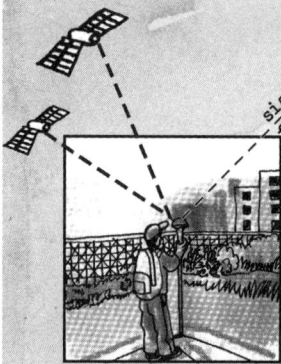

SORRY, I WANT TO RECORD ONE MORE POINT . . . UH . . . EK EXTRA POINT . . . UH . . . NEEDING.

SUN HAI . . . VERY HOT HAI. [IT IS SUNNY AND HOT]

YOU ARE OKAY?

HAAN! OKAY, OKAY! [YES, I'M OKAY!]

ARREY MUMBAI MAIN BOHOT SARA FIRST CLASS POINTS HAI! AAPKO MAIN SARA POINTS DIKHAYEGA AUR USKE BAAD AAPKO IDHARICH REHNE KA MANN KAREGA! CHENNAI KO TA-TA BYE-BYE! [ARREY MUMBAI HAS MANY FIRST CLASS POINTS! I'LL TAKE YOU TO ALL OF THEM AND AFTER THAT YOU WON'T WANT TO LEAVE! TA-TA BYE-BYE TO CHENNAI!]

(TO ME): TYALA KAHICH SAMAJHALE NAHI NA? HEHE . . . (TO ME): [HE DIDN'T UNDERSTAND ANYTHING, DID HE? HEHE . . .]

THE SURVEYORS CHOOSE SITES THAT ARE EASILY DISTINGUISHABLE ON THE SATELLITE IMAGE.

THEY ALSO MAINTAIN RECORDS OF WHERE THE READINGS WERE TAKEN, TIMING, AND SITE CONDITIONS.

height

serial number

reference image (location on the satellite image)

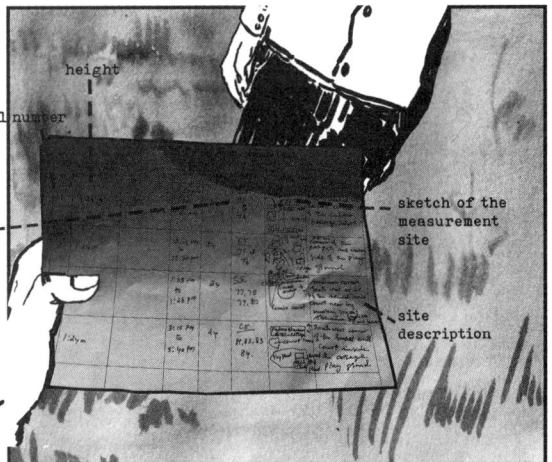

sketch of the measurement site

site description

TO DRAW THE CZMP USING GIS SOFTWARE, THE TEAM COMBINES THE SURVEY DATA WITH HISTORICAL RECORDS, WHICH PROVIDE ADDITIONAL INFORMATION SUCH AS DISTRICT AND PLOT BOUNDARIES.

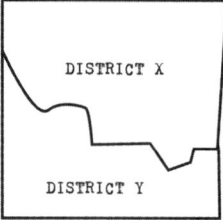

DISTRICT X

DISTRICT Y

DEPENDING ON THE MATERIAL CONDITION OF THESE DOCUMENTS, THEY ARE SCANNED, PROCESSED, AND DIGITALLY OVERLAID ON THE BASE SATELLITE IMAGE.

ONCE GEO-REFERENCED, THEY ARE TRACED AND ADDED IN AS A PART OF THE CZMP.

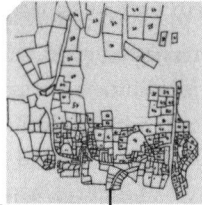

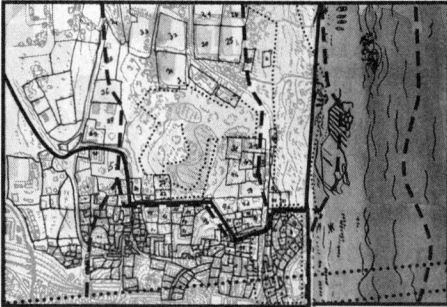

IMAGE IS PROCESSED

GEO-REFERENCED ONTO THE SATELLITE IMAGE

DATA FROM THESE RECORDS ARE DIGITALLY TRACED

HIGH TIDE LINE

CRZ I (MANGROVES)   CRZ III   BUFFER ZONE   LOW TIDE LINE

DATA FROM THE COASTAL SURVEYS, INCLUDING THE HTL, THE LOCATION OF ECO-SENSITIVE ZONES, AND EXISTING INFRASTRUCTURE, ARE ADDED.

CRZ II

CRZ BOUNDARY (500 m FROM HTL)

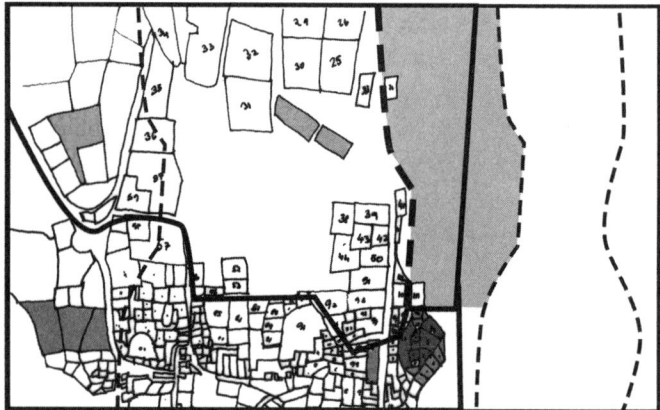

Surveyors like Anbu deploy a range of strategies to reduce error, and these involve a number of subjective choices.[9] Apart from carefully selecting control points, Anbu also kept meticulous records, which he maintained in a large booklet he carried. As the municipal worker held the GPS device, Anbu would juggle the reference map and the booklet in which he would record the height at which the reading was taken, the time and duration of the GPS reading, and a short description and a diagram of the location that would allow him or any other surveyor to recognize the point at a later date. He explained that since the CZMP project is conducted by a group, it was important to maintain a detailed log of all points in a manner identifiable by other surveyors. If, in the future, there was any doubt about a surveyed point, any team member could look up this log to avoid potential errors.

Just as there are specific practices of producing accuracy in the field—selecting points with care, calibrating instruments, and waiting to ensure it gets enough data to eliminate errors and keeping meticulous records—there are measures that ensure accuracy in the act of drawing and compositing the CZMP in a GIS software. To draw the CZMP, the surveyors must gather data from different sources, including other older surveys drawn at different scales. The surveyors work not only around data errors, but also navigate different drawing conventions, and errors that are produced because of the material qualities of documents. For instance, in a manual on creating CZMPs for a district in Kerala, the authors note that the cadastral surveys are often only available as paper copies. Some of these copies are extremely old, which means that the surveyors have to look up changes to the plots in that cadastral sheet. Other times, cadastral surveys are only available as photocopies (often photocopied several times over).[10] As the photocopying process often results in distortions and loss of data, the surveyors have to ensure that they correct for those errors in the process of putting together the several layers of information that eventually become the CZMP.

In a book on practices of technical representation, Katherine Henderson (1998) writes that drawing allows engineers to engage in multiple forms of knowledge simultaneously. They "serve as the holding ground where codified and uncodified knowledge can meet," where the experiential, kinesthetic, mathematical, archival, and visual knowledge can mix (Henderson 1998, 12). The process of producing the CZMP is such a meeting ground for forms of knowledge and mixing practices. It allows multiple engineers, surveyors, and government officials to come together to render the coast through information collected on different kinds of interfaces—booklets, sheets, databases, on instruments, and on geographic interfaces. At each stage in the process, the surveyors take care to reduce interpretive flexibility as much as possible—that is, to take, record, and convert the information in ways that effaces the author. And yet, each of these moves involves choices based on different sorts of visual expertise.

A Cartographic Canvas

The reasons for how plans came to be at the center of the CRZ has to do with a combination of factors, including the changing nature of democratic politics, the shifting frameworks of environmental management, and technological advancement. It is also the outcome of a history of disasters—events that left their mark on policies and reframed the managerial strategies of governing the coast. The CZMPs are the product of all these factors—policies, technology, management discourses—and they imposed frameworks of power based on systems of controlling land, which turned the coast into a territory for capital accumulation.[11]

The origin story of the CRZ policy is that in 1981, after Prime Minister Indira Gandhi visited Puri, a city in the state of Orissa, she issued a letter to the chief ministers of India's coastal states asking them to establish a five-hundred-meter buffer zone to protect the shoreline from degradation (Sahu 2016; Sundar 2014). This prime ministerial directive eventually became the foundation of the 1991 CRZ. One of the main problems of the 1991 CRZ was its regulatory model, which imposed a uniform set of rules for a diverse coastline. The 2011 CRZ was supposed to respond to this problem by making greater room for participation by stakeholders and giving greater power to the state governments. The higher-resolution CZMPs were a crucial lynchpin in this shift as they were supposed to reflect nuances on ground.

This shift from a regulatory to a more decentralized coastal model followed shifting trends in environmental governance, which, K. Sivaramakrishnan (Sivaramakrishnan 2011) writes, is an outcome of the changing nature of democratic politics in India, particularly, the distribution of power between the center and the states. Sivaramakrishnan shows how over the last few decades the legal apparatus has, through various acts, constitutional amendments, and judicial actions and judgments, moved from a model where the environment was largely regulated by a central authority to one that now includes a number of state and nonstate actors operating at different levels. These shifts parallel changes in environmental management discourses as well: from a central authority that sets the mandate, executes plans, and distributes resources, it has now moved to a model based on caretaking conducted through hybrid organizational structures.

Digital technology is an important aspect of this caretaking process: recent policies position technology as a means of translating ideas of management into bureaucratic terms. In the case of the CRZ, technology—particularly survey technology—is fundamental to coastal management as a governance practice, as it determines the elements that have conservation and resource value and the distribution of rights and resources to coastal communities. These survey technologies are entangled in a web of questions such as, Whose responsibility is it to care? Who are the managers, stewards, owners, and wards? And what is the coast as the object of care? The evolution of environmental policies and survey technologies were not two separate stories that intersected from time to time when technology was required to "enact" policy. The role that survey technology played was far more crucial—advances in survey technology and satellite mapping changed ideas of responsibility, rights, and forms of control in those policies.

In India, a comprehensive environmental policy was not put in place until 1986 when the central government, under great pressure due to the Bhopal gas disaster, passed the Environment Protection Act (EPA) (Abraham and Abraham 1991). Beside the event of the disaster, the 1986 EPA also sought to mitigate some of the damage caused by the excessive use of fertilizers, pesticides, and high-input intensive technologies during the Green Revolution, which was crucial for establishing food security in the decades following independence (Vyas and Reddy 1998) and to mitigate the pressures of rapid growth and uneven development across different regions (Bowonder 1986). The EPA (1986) is an important part of the CRZ's history because it gave the central government authority to pass laws to protect any part of the environment.[12] Because of the power granted by the EPA, the Ministry of Environments and Forests and Climate Change (MoEFCC) became (and still is) the central body that established boards that oversee things such as on waste management and air pollution, and which put project clearance mechanisms in place. The MoEFCC, under the provisions of the 1986 EPA, passed the first CRZ notification in 1991, and all other subsequent notifications.

Historian Sheila Jasanoff writes that critics saw this arrangement—of the center passing environmental laws that affected the entire country—as perpetuating "neo-colonial relation[s] between the citizen and the state, effectively denying the former any participatory rights in the management of the country's natural resources" (Jasanoff 1993). The 1991 CRZ faced this criticism when it was released—it was regarded as another instance of the center's imposition of authority, and was criticized for its top-down structure. Advocates for fishers' rights wrote that while it was necessary to create a framework for coastal management and establish a state-level authority to monitor it, it was also necessary to involve stakeholders such as communities that lived in and relied on the coast for their livelihood (Equations 2007). To do this, the CRZ had to be decentralized and made to work beyond the district level and engage local government. Moreover, the CZMPs made under the 1991 CRZ mapped the coast at the district level, leaving out municipalities and *panchayats*—institutions, which typically have a better understanding of the relations on the ground (Sharma 1996). Thus, the growing power of local institutions was one important reason why the coast had to be mapped at a higher resolution.

With the rising popularity of participatory models, the criticism of the 1991 CRZ as a top-down enterprise gained greater force. Movements such as the Kerala People's Campaign for Decentralized Planning fueled the opposition against a central monitoring authority (Isaac and Franke 2002). Moreover, in 1993, the central government passed the Seventy-third and Seventy-fourth constitutional amendments, which transformed the former two-tiered system (central and state governments) to a "three-tiered" system that gave local authorities like municipal corporations (urban) and panchayats (rural) greater power in the planning process (Menon and Kohli 2008; Rao and Ramana 2008). In the two decades between the CRZ notifications, decentralized planning mechanisms emerged that routed governance through local institutions. This move was accompanied by economic liberalization and from the late 1990s, environmental planning policies and interventions were critically shaped by global capital flows and became increasingly market oriented (Mansfield 2004). Conservation policies that followed were characterized by a rescaling of the state's power and what Shalini Randeria (Randeria 2003) has called the "hybridization of law"—where international standards, organizational structures, concepts of the environment, and relationships between public and private actors are reframed in relation to a financial landscape shaped by transnational flows of capital. The national policies passed in this period repackaged the environment as a commodity through expanded schemes for ecotourism initiatives, by basing conservation strategies on charismatic species like tigers and elephants, or by focusing on particular high-value crops (Münster and Münster 2012). This decentralization process set up the current hierarchical arrangement that characterizes coastal management where the central government still functions as the apex authority under which the state-level Coastal Zone Management Authorities (CZMAs) function, with NGOS, as local institutions working on ground to ensure community participation.

Coastal governance also changed because of international events like the United Nations Conference on Environment and Development that took place in 1992 (and other subsequent international agreements), which promoted Integrated Coastal Zone Management (ICZM) techniques. ICZM, which is the backbone of the 2011 CRZ policy, frames the coast as a dynamic zone where topography, ecology, land use, population, and economy exist in a state of constant flux (Bowen and Riley 2003). While this idea of the coast as a dynamic collective might seem like a good thing where conservation is concerned, coastal management when largely framed in relation to developmental goals, results in degradation (Sharma 1996). This focus on developmental timelines combined with changing environmental governance fostered public-private partnerships, granting regulatory powers to quasi-state entities such as NGOs, international organizations, and hybrid regulatory groups. Even though the ICZM conceptualizes the coast as mixture of several dynamic elements, the way coastal ecological policies are framed makes it abundantly clear that not all elements of this dynamic mix count or are regarded as equal. Indeed, the focus of ICZM-based policies has emphasized the use value of certain marine or coastal resources and based policymaking on the effective consumption of these resources (Bavington 2011).

The change from a regulatory framework to one based on management impacted the role that surveys played in the CRZ policy. The 1991 CRZ is a slim document of a few pages and it mostly focuses on listing activities prohibited in the coastal zone. In contrast, the 2011 CRZ runs into several pages as the CZMPs were expanded to include revenue sources such as small- and large-scale fisheries, boat-landing sites and jetties, and other fisheries infrastructures such as refrigeration units. For example, in the 2011 CRZ, the coastal states were also directed to record and identify areas with resource potential such as ecotourism zones, greatly expanding the list of things to be identified, mapped, and fixed on the plans. Thus, with changing institutional arrangements that pulled in local organizations and changing ideas of care and management, the number of elements to be mapped drastically expanded, creating the need for higher-resolution maps on which a greater number of elements could be located, identified, and reclassified.

The 1991 CRZ stipulated that the CZMPs use topographic survey sheets as base maps. Topographic surveys are maps where the primary objective is to record the physical features of a terrain, and in India, these maps are generally drawn at the scale of 1:25,000 or 1:50,000. At this scale, maps only show detail at the district level, and they do not include individual villages or property boundaries. The 2011 CRZ stipulated that the CZMPs would keep the topographic scale, but also include a set of large-scale maps, drawn at 1:4,000. These maps would use cadastral sheets—surveys that maintain a record of individual properties for purposes of revenue and taxation calculations. While this move to include details at the cadastral scale meant that there would be greater clarity regarding individual properties, the ability to identify resources in greater detail, and aid in making decisions at the level of local institutions, it further entrenched the coast into systems of governing and controlling land.

Apart from identifying ecological commodities, flora, fauna, and topographic formations also began to be recognized as valuable for their role in mitigating natural disasters. After the 1991 notification, the MoEFCC commissioned several research committees to investigate different parts of the CRZ notification based on which it published amendments. In late 2004, as the Swaminathan Committee was working on its report, a tsunami triggered by an earthquake in the Indian Ocean claimed more than ten thousand lives in India. As the eastern coast was reeling from this event, in the following year, Maharashtra state was hit by severe floods during the monsoons. Both these events, plus a super cyclone that had struck the east coast in 1999, left deep marks on the CRZ notification and were critical in shaping its present form and its mapping process (Kesavan and Swaminathan 2007; Swaminathan et al. 2009).[13]

In the introductory paragraphs of its first review report of the CRZ, the Swaminathan Committee characterizes the coast as an ecologically rich site under threat from both anthropogenic pressure and from natural disasters. But from this report, it was clear that danger has directionality—it comes from the sea and inflicts damage on the land, damage that is largely measured in terms of human lives and property. It noted that there was a need to relocate coastal communities to safer areas, to protect coastal features such as mangroves, coral reefs, and plantations that had greatly reduced the damage caused by tidal waves and cyclones. It advocated protecting these coastal elements and installing warning systems to minimize the threat to human life and to property (Swaminathan et al. 2009). The Swaminathan Committee's report follows observations made in several articles written by academics, activists, and journalists suggesting that the outcome of events like the tsunami or the cyclone might have been different had the government taken greater care in enforcing the CRZ notification and in conserving ecological elements like sand dunes and mangrove forests.

After the 2005 flood in Mumbai, there was widespread speculation about whether the severity of the flood could have been controlled had the city's "natural drainage" channels such as rivers, mangroves, and salt pans been mapped, demarcated, and protected under policies like the CRZ (CAT, Conservation Action Trust 2006). Similarly, after the 2004 Indian Ocean tsunami, several studies used remote sensing technology and field maps to compare the impact of the tsunami wave in different regions that had varying land features and vegetation (Bhalla 2007; Olwig et al. 2007). For instance, a study by Mascarenhas and Jayakumar (2008) along the coast of Tamil Nadu in South India, surveyed different sections of the coast to examine the effect of the tidal wave on different flora and land formations along the southern coast. By correlating the vegetation and the damage, the authors recommended planting vegetation and emphasized the need to implement the CRZ notification. Such post-tsunami studies advocated the use of high-resolution mapping to survey, study, and identify ecologically important species and formations. They were able to show the capacity of vegetation and dunes to act as bio-shields (Bhalla 2007; Mukherjee et al. 2010). The studies quickly made their way from scholarly circles into newspaper articles via publicly accessible reports, and subsequently, ecological formations too became a part of the expanding list of things to be identified and located in the CZMPs. Consequently, in the aftermath of these events, the CRZ policy began identifying and locating an expanding list of plants, animals, and landscape formations.

Because of disasters, changing governance practices, arrangements, management discourses, and changing technologies, the CRZ changed a lot in the two decades between the 1991 and the 2011 notifications. The coastal plans became more important to the CRZ as the policy evolved. They were the core legal instruments and material objects that arranged the CRZ as a bureaucratic entity and transformed the coast as an object of governance. However, their capacity to remake the coast was directly proportional to their capacity to present a highly detailed representation of it, that is, to their visual resolution. The plans drew on the power of increasingly sophisticated satellite technology. Whereas earlier, the CRZ delimited large swathes, by 2011 it zoomed in to show individual properties, community lands, coastal commons, resources, plantations and groves, ecotourism spots, and features such as dunes within the plan. This contributed to an increasingly fragmented understanding of the coast and further entrenched coastal politics in terms of claims over parcels of the coast.

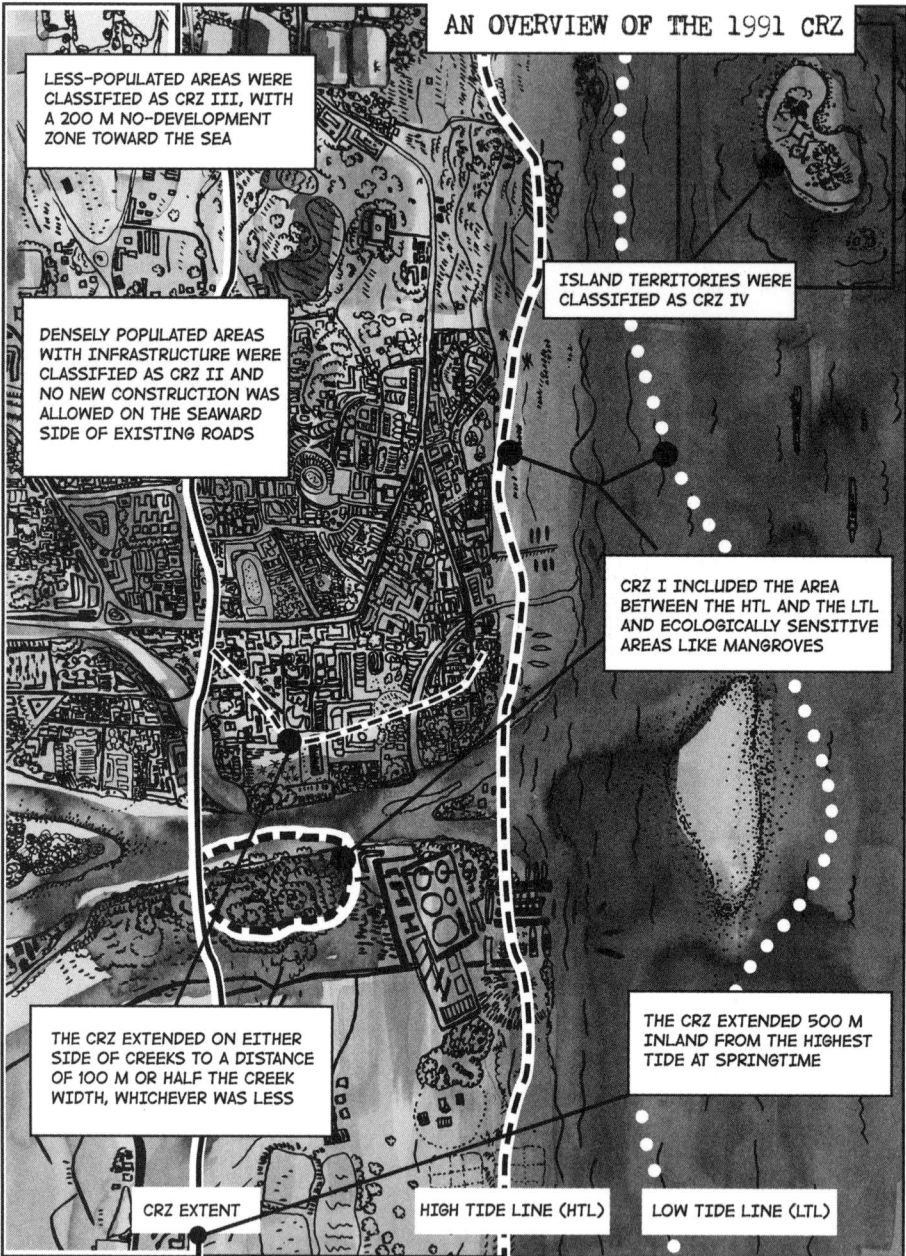

AN OVERVIEW OF THE 1991 CRZ

LESS-POPULATED AREAS WERE CLASSIFIED AS CRZ III, WITH A 200 M NO-DEVELOPMENT ZONE TOWARD THE SEA

ISLAND TERRITORIES WERE CLASSIFIED AS CRZ IV

DENSELY POPULATED AREAS WITH INFRASTRUCTURE WERE CLASSIFIED AS CRZ II AND NO NEW CONSTRUCTION WAS ALLOWED ON THE SEAWARD SIDE OF EXISTING ROADS

CRZ I INCLUDED THE AREA BETWEEN THE HTL AND THE LTL AND ECOLOGICALLY SENSITIVE AREAS LIKE MANGROVES

THE CRZ EXTENDED ON EITHER SIDE OF CREEKS TO A DISTANCE OF 100 M OR HALF THE CREEK WIDTH, WHICHEVER WAS LESS

THE CRZ EXTENDED 500 M INLAND FROM THE HIGHEST TIDE AT SPRINGTIME

CRZ EXTENT   HIGH TIDE LINE (HTL)   LOW TIDE LINE (LTL)

THE COASTAL POLICY EVOLVED HAND IN GLOVE WITH THE DEVELOPMENT OF SATELLITE TECHNOLOGY AND GIS. THE FIRST VERSION OF THE CRZ POLICY WAS RELEASED IN 1991 AND IT DIVIDED THE COAST INTO FOUR BROAD ZONES OF VARYING ECOLOGICAL SENSITIVITY, AND USED MAPS DRAWN AT THE REGIONAL SCALE. THE 1991 CRZ WAS A SLIM DOCUMENT OF A FEW PAGES, AND IT FOCUSED ON ACTIVITIES PROHIBITED IN THE COASTAL ZONE.

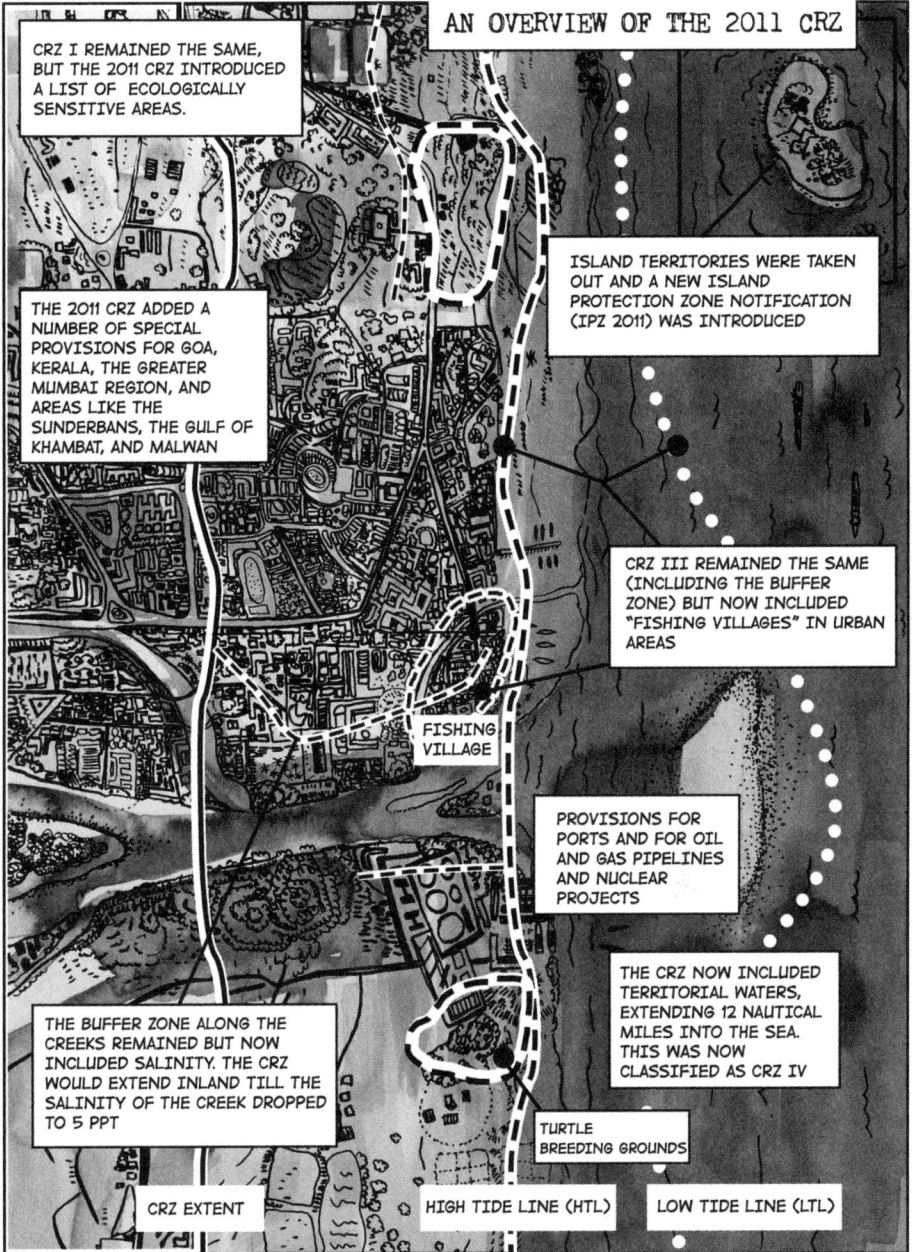

AN OVERVIEW OF THE 2011 CRZ

CRZ I REMAINED THE SAME, BUT THE 2011 CRZ INTRODUCED A LIST OF ECOLOGICALLY SENSITIVE AREAS.

ISLAND TERRITORIES WERE TAKEN OUT AND A NEW ISLAND PROTECTION ZONE NOTIFICATION (IPZ 2011) WAS INTRODUCED

THE 2011 CRZ ADDED A NUMBER OF SPECIAL PROVISIONS FOR GOA, KERALA, THE GREATER MUMBAI REGION, AND AREAS LIKE THE SUNDERBANS, THE GULF OF KHAMBAT, AND MALWAN

CRZ III REMAINED THE SAME (INCLUDING THE BUFFER ZONE) BUT NOW INCLUDED "FISHING VILLAGES" IN URBAN AREAS

FISHING VILLAGE

PROVISIONS FOR PORTS AND FOR OIL AND GAS PIPELINES AND NUCLEAR PROJECTS

THE CRZ NOW INCLUDED TERRITORIAL WATERS, EXTENDING 12 NAUTICAL MILES INTO THE SEA. THIS WAS NOW CLASSIFIED AS CRZ IV

THE BUFFER ZONE ALONG THE CREEKS REMAINED BUT NOW INCLUDED SALINITY. THE CRZ WOULD EXTEND INLAND TILL THE SALINITY OF THE CREEK DROPPED TO 5 PPT

TURTLE BREEDING GROUNDS

CRZ EXTENT

HIGH TIDE LINE (HTL)

LOW TIDE LINE (LTL)

THE SECOND VERSION OF THE CRZ WAS RELEASED IN 2011, AND THOUGH IT KEPT ITS PREDECESSOR'S STRUCTURE, IT RELIED HEAVILY ON SATELLITE AND SURVEY TECHNOLOGIES TO PRODUCE A HIGH-RESOLUTION PICTURE OF THE COAST. IT EXPANDED THE LIST OF PERMISSIBLE ACTIVITIES AND PROVIDED SPECIAL PERMISSIONS FOR LARGE-SCALE INFRASTRUCTURE PROJECTS. IT INCLUDED HIGH-RESOLUTION MAPS OF INDIVIDUAL PROPERTIES AND PROVISIONS FOR DISTRIBUTING LAND RIGHTS TO COASTAL COMMUNITIES. THUS, THE NUMBER OF THINGS TO BE MAPPED AND THE LEVEL OF DETAIL INCREASED EXPONENTIALLY.

AN OVERVIEW OF THE 2018 CRZ

THE RECENTLY RELEASED 2018 CRZ FURTHER ATOMIZES THE COAST AS IT INTRODUCED SEVERAL NEW SUBZONES AND RELAXED DEVELOPMENT RESTRICTIONS.

CRZ I

CRZ I-A: ECO-SENSITIVE AREAS

CRZ I-B: INTERTIDAL AREAS

CRZ II

AREAS THAT ARE ALREADY DEVELOPED CLOSE TO THE WATER

CRZ III

CRZ III-A: AREAS WITH A POPULATION DENSITY HIGHER THAN 2,161 PEOPLE/ SQ KM. THE NO-DEVELOPMENT BUFFER ZONE REDUCED TO 50 M

CRZ III-B: AREAS WITH POPULATION DENSITY LOWER THAN 2,161 PEOPLE/ SQ KM (AS PER THE 2011 CENSUS). BUFFER ZONE REMAINS AT 200 M

CRZ IV

CRZ IV-A: 12 NM FROM THE LTL

CRZ IV-B: TIDALLY INFLUENCED WATER BODIES

APART FROM THESE NEW SUBZONES, THE 2018 CRZ GRANTED SPECIAL PERMISSIONS FOR BUILDING MONUMENTS AND ALLOWED TOURISM PROJECTS IN ECOLOGICALLY SENSITIVE ZONES. IT ALSO REMOVED RESTRICTIONS ON INFRASTRUCTURES TO PROMOTE THE GOVERNMENT'S PORT DEVELOPMENT PROGRAM. THE 2018 CRZ WAS AMENDED WITHOUT ANY PUBLIC CONSULTATION.

THE 2018 CRZ WAS BASED ON RECOMMENDATIONS OF THE SAILESH NAYAK COMMITTEE, WHICH WAS CONSTITUTED IN 2015. THE NAYAK COMMITTEE FOCUSED ON THE COAST'S ECONOMIC POTENTIAL.

A 2002 ARTICLE AUTHORED BY THE COMMITTEE HEAD, SAILESH NAYAK, A SCIENTIST WITH EXPERTISE IN REMOTE SENSING, PRESAGES THE 2018 CRZ. IN IT, NAYAK NOTES THAT THOUGH THE COAST IS AN ENVIRONMENTALLY FRAGILE ZONE, IT CAN BE SUSTAINABLY DEVELOPED FOR ECONOMIC GROWTH. REMOTE SENSING AND HIGH-RESOLUTION SURVEYS ARE, NAYAK WRITES, IMPORTANT TOOLS TO ENSURE SUSTAINABLE DEVELOPMENT AS THEY PROVIDE AN ACCURATE PICTURE OF THE COAST.

NAYAK'S (2002) ARTICLE PROVIDES DETAILED GUIDELINES FOR MAPPING COASTAL LAND-USE TO IDENTIFY ELEMENTS OF ECOLOGICAL AND RESOURCE VALUE.

Producing Accuracy, Evidence, and Authority

During the thirty-first meeting of the National Coastal Zone Management Authority (NCZMA), which was held in 2016, representatives from coastal states were asked to give an account of the progress they had made on their CZMPs.[14] In a lengthy response, the representative from Maharashtra said that the state had divided the task of creating the plans by individual districts, which were in the process of preparing both the small- (1:25,000) and the large- (1:4,000) scale drawings. These surveys would be used to produce digital base maps with details including property numbers and the area of individual plots. Ecologically sensitive areas such as mangroves too would be demarcated along with fishing infrastructure, fishing villages, and points of high and low coastal erosion. The representative stated that the MCZMA's objective was to produce a "GIS based coastal information system" that would "work as an information kiosk," or interface, through which individuals could check whether their property was a part of the CRZ and the kinds of restrictions that would come into play if it was.

While no such interface exists, the MCZMA's idea for a kiosk is descriptive of the wide reach of GIS-based systems and the growing use of digital technologies in governance in India. Paper maps and digital interfaces produce completely different senses of spatial immersion and navigation: digital interfaces are thought to embed people into the landscape in intimate ways, because unlike paper maps where the viewer's eye is situated outside the frame, digital systems put individual viewers at the center of a map that moves with the eye (November et al. 2010). The way in which a GIS-based interface moves seamlessly between scales—particularly, zooms into a landscape—confers a sense of accuracy, immersion, and immediacy to the map, a sentiment that lends itself well to the idea of a digital bureaucracy that is transparent, self-evident, seamless, and accessible to all (Crampton 2009). This accuracy is thought to be an inherent property of the map image and its resolution: the higher the map resolution, the greater the level of detail, and greater the capacity for the map to speak spatial truths. However, the process of producing the CZMPS and the ways in which maps are called into evidence in court cases shows that accuracy and its relation to spatial truths is complex.

The problem of producing accurate CZMPs has long haunted the CRZ. The problem itself has taken on different forms—which is to say that the problem of accuracy in the 1991 CRZ was not the same as the ways in which accuracy was dealt with in the 2011 CRZ. Rather than the property of an image or a technological outcome, the CZMP shows a different kind of accuracy at work—one that is caught up in questions of authenticity, cross-referencing, materiality, and is actively produced through bureaucratic processes and through drawing practices.

In the early versions of the 1991 CRZ notification, the government had not specified any method, scale, or mapping criteria for identifying the coastal zones. It was only in 1994, in an amendment published by the MoEFCC, that a directive was issued to the state governments to prepare CZMPs based on maps approved by the Survey of India (SoI), within a period of one year, and in accordance with the guidelines specified in the notification. However, despite MoEFCC's order, it was not until 1996 that preliminary plans were produced by each of the states. These plans were prepared only after the Supreme Court of India passed a judgment in a public interest litigation case (PIL), *Indian Council for Enviro-Legal Action v. Union of India* in that year.[15] The petitioner in the case, an NGO working environmental issues, argued that the Maharashtra state's failure to produce a CZMP had led to unregulated development, causing rapid coastal degradation. The petitioners pointed out that the Maharashtra government had allowed polluting industries as result of failing to prepare land-use surveys of the region that would have provided a complete record of all the ecologically important areas. Siding with the petitioner, the court, in its judgment, called for a proper and detailed implementation of the law through the creation of CZMPS.

In the historic timeline of the CRZ policy and its development, this case stands out as it marks one of the first instances where a higher court issued a mandamus to state authorities. However, this judgment must be understood in the context of a judiciary system that took on an increasingly proactive role in relation to cases pertaining to the environment through the instrument of the public interest litigation (PIL). PILs, which were developed to benefit the socially and economically poor public, are instrumental in environmental activism and protection in India. PILs, Sivaramakrishnan (2011, 910) writes, became instruments that courts could use to adopt an "activist persona to focus on socio-economic rights and the quality of life." Beginning with orders such as the one issued by the Supreme Court in 1996, there were several crucial developments in the CRZ that were engineered through judgments passed in various high courts throughout the country. The way maps are presented as evidence in these examples (in both PILs and cases concerning private property) casts light on some aspects of the relation between image resolution and authority.

As a result of the 1996 order, all the coastal states submitted sets of CZMPs to the MoEFCC, which were made using survey sheets approved by the SoI.[16] As a base, these plans used topographic survey sheets produced by the SoI at the scale of 1:25,000 on which the zones were plotted using data from the IRS-IA remote sensing satellite.[17] Since the early CRZ maps used the topographic scale, which is more standard for depicting terrains over large regions, it was very difficult to tell whether individual properties located on the edge of the zone fell under the CRZ.[18] For example, in 2003, a case, *Citizens Interest Agency vs. Lakeshore Hospital*

2003 (3) KLT 424, concerning a hospital that was allegedly built in the CRZ, was brought to the Kerala High Court. The petitioner stated that the structure, Lakeshore Hospital, should not exist on the plot specified as it was in an ecologically sensitive zone, categorized as CRZ I. This category concerns protected forests and ecologically sensitive zones, and there are several developmental restrictions in place. In the judgment, the chief justice, J. L. Gupta, noted that while environmental clearances had been obtained for constructing the hospital, it was difficult to ascertain whether the hospital was in violation of the CRZ regulation as the CZMPs used by the management authority were drawn from enlarged topographic survey sheets, which showed landscape features and not property details. The judgment referred to a letter prepared by the MoEFCC directing the Kerala government to prepare local level CZMPs at the cadastral scale, which would show these property details.[19] The case was dismissed on the grounds that the petition was premature until a map was drawn at a scale large enough to clear the ambiguity regarding individual properties.[20]

While such cases set a precedent for high-resolution maps, the inability to see individual parcels and properties was just one part of the problem. As other cases show, it was also a problem of locating the CRZ on maps, plans, and surveys that had been approved by specific state institutions, thus placing the CRZ within the continuity of previously recognized and legitimate governmental apparatuses. This is apparent in the cases where topographic surveys were pitted against satellite imagery from sources such as Google Earth, which had become publicly available in 2004. For instance, in *Ansari Kannoth, 'Santhwanam' vs State of Kerala (WP(C). No.12623 of 2010(S))*, the petitioner filed a PIL against the state for operating a park that they claimed was damaging the existing mangrove forest in the area. The petitioner also argued that the CZMPs drawn using topographic sheets (submitted as evidence by the state) were incorrect as they did not specify the area the sheet represented, and the park's location on the map could not be identified.

The Kerala State Coastal Zone Management Authority (KSCZMA) countered that its maps had been approved by the MoEFCC. At this juncture, the state government also produced Google Earth images to claim that in 2003, when the park was built, there were no mangroves on site, and therefore, the theme park could not have violated any CRZ regulation as without the mangroves, the area could not be categorized as a CRZ I zone. Ruling against the KSCZMA, the chief justice declared that Google Earth maps could not overrule the topographic sheets for two reasons: the first was that the satellite image could not be regarded as an accurate depiction as there was a possibility that cloud formations could obscure details on ground or that the mangrove could have come up after 2003. Second, the topographic sheets had the backing of the state as they had been verified by the SoI. Moreover, since there was no reason why mangroves planted after 2003 could not fall in the CRZ I category, the case was ruled in favor of the petitioner.

Both the examples grapple with the issue of presenting maps as evidence. While the first case concerns the production of high-resolution maps to verify whether a given property falls within the CRZ, the second case shows that, though important, spatial truths do not rest entirely on the ability to see detail. The case shows that the point of preparing high-resolution maps is not just to reinforce the relation between cartographic representation and geographic realities, rather, it is about what kind of cartographic representations can come to serve as evidence. Not only does the map in question have to *show* the plot or property concerned (hence the need for a higher-resolution image), it also has to be a map that is recognized by the state or bear the mark of a state official authorized to publish and distribute the survey sheet.[21] Similarly, though the Google Earth images provide visual details of the areas in question, these are often not accepted as evidence because do not carry any marker of approval, legitimacy, or sanction from a state institution. As a result, to arrive at a decision, courts try to trace the stamps, notes, and other marks that a map accumulates in its passage through a bureaucratic institution.[22] The spatial claims that maps make gain greater force when they are put in relation to existing survey records. In other words, it is not enough that a map records the landscape in detail, but it also must establish its authority through official recognition.

In an article on missing planning maps and cartographic actions in Ho Chi Minh City, Erik Harms (2020) writes that there is nothing automatic about the way that maps tell the truth or perform accuracy. Harms writes that after all, maps are drawn by people toward specific ends, and this means that the question of whether a map is accurate is not as important as what that accuracy does. The CZMPs produced when the first CRZ notification was passed were deemed inaccurate because of their low resolution. To resolve this problem, the 2011 CRZ stated that they would be drawn at the cadastral scale, which would clear up ambiguities associated with property development. However, this new drawing project was not just in aid of making accurate maps; it was also a way of exerting greater authority on the coast to recast it within the state's developmental visions. Drawing the CZMPs at the higher, cadastral scale played a crucial role in this process.

The CZMPs are not just drawn on a blank page; they use existing surveys as base maps. Thus, the CRZ comes to life not just through act of inscribing lines on paper, but by inscribing it in relation to other documentary regimes.[23] While the 1991 CRZ used topographic sheets published by the SoI as base maps, the 2011 CRZ used cadastral survey sheets. Unlike topographic sheets, which cover large regions, cadastral sheets are used as building blocks in land administration and taxation. The cadastral maps are linked to land registers that track use, development, and ownership status. That the CRZ used the cadastral survey for creating a more detailed map of the coast is not arbitrary. From the early modern period onward, the cadastral map has functioned as an instrument to establish rule of property, extend the reach of the state and its capitalist structures (Kain and Baigent 1992).

In the Indian context, cadastral surveys, as Edney (1997) writes, were an important part of the British Empire's fantasy of making a perfectly detailed map of the empire even if they mapped territory at a scale that was too large to comprehend as a unified image. These surveys were generally mapped at the scale of sixteen inches to the mile (1:3960) or as was the case in some parts, at much larger scales that showed details such as individual trees (Edney 1997). In calling the cadastral surveys into service, the CRZ repeated the same fantasy project of creating a detailed map that would consolidate the coast as a territorial unit, while legitimizing it by linking it to existing documents.

IN LATE 2011, I DESCENDED INTO THE DARK UNDERPASS CONNECTING THE CHHATRAPATI SHIVAJI TERMINUS WITH THE MCGM BUILDING. THE UNDERPASS WAS WARM FROM THE HEAT OF THE CROWD AND DEV KOLI, A REPRESENTATIVE FROM MY FIELD SITE, MALVANI KOLIWADA, WAS WAITING FOR ME. WE MADE OUR WAY UP THE STAIRS AND WERE USHERED INTO A CONFERENCE ROOM FOR A PUBLIC MEETING ON SETTING UP SEWAGE TREATMENT PLANTS AT THE COAST.

THERE WERE ABOUT 20 PEOPLE IN THE ROOM, ALMOST ALL MEN FROM THE CITY'S FISHER COMMUNITY. THE MUNICIPAL OFFICIAL TURNED ON THE PROJECTOR AND TOOK US THROUGH A SLIDESHOW THAT EXPLAINED THAT UNDER THE NEWLY RELEASED COASTAL REGULATORY ZONE POLICY (2011 CRZ) THE CITY COULD NOT RELEASE UNTREATED WASTE INTO THE SEA.

THE NEXT SLIDE HAD A DIAGRAM THAT SHOWED HOW THE MCGM WOULD CONSTRUCT A MAIN LINE THAT WOULD LEAD INTO A TREATMENT FACILITY AT THE CITY'S EDGE. FROM THERE, PIPES WOULD CARRY THE WASTE FAR OUT INTO THE SEA WHERE IT WOULD BE ALLOWED TO DIFFUSE WITH THE SALTY WATER. THE OFFICIAL CONCLUDED BY SAYING THAT THIS WAS HOW SEWAGE TREATMENT WORKED ALL OVER THE WORLD AND THAT IN THE LONG TERM IT WOULD BE TO THE FISHER COMMUNITY'S BENEFIT. SURELY, HE SAID, THEY COULD SEE THE SENSE IN REDUCING COASTAL POLLUTION.

SARCASTIC LAUGHTER AND NODS RIPPLED THROUGH THE AUDIENCE WHEN SOMEONE RESPONDED BY SAYING THAT OF COURSE THEY KNEW ABOUT POLLUTION—IT SHOWED UP EVERY DAY IN THEIR NETS. ON MANY DAYS, WHEN THE NETS WERE HAULED IN, EVERYTHING WAS COVERED IN PLASTIC. THE FISHERMAN POINTED OUT THAT THEY WERE NOT AGAINST THE PLAN IN THEORY. BUT THEY WERE THE MOST AFFECTED BY THE POLLUTION, HE SAID, BECAUSE THE PIPES WOULD EXIT AT THE SPOT WHERE THE NEAR-SHORE FISHER COMMUNITY CAST ITS NETS.

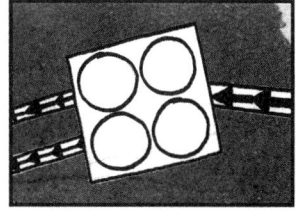

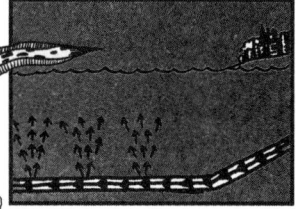

IN THE BACK AND FORTH, IT WAS CLEAR THAT THE MUNICIPAL OFFICIALS DID NOT KNOW THE WAY IN WHICH NEAR-SHORE MARINE TENURE WORKED. THE OFFICIALS IMAGINED THAT THE FISHER COMMUNITY COULD SIMPLY MOVE AWAY FROM THE LOCATION OF THE PIPES. HOWEVER, THE NEAR-SHORE WATERS OF MUMBAI ARE DIVIDED AND THERE ARE SPECIFIC SPOTS WHERE EACH COMMUNITY FISHES.

THE MEETING ENDED WITHOUT A RESOLUTION. AS DEV KOLI AND I WALKED TOWARD CHURCHGATE STATION, HE SAID THAT THIS WAS PRETTY MUCH HOW EVERY MEETING REGARDING THE CRZ WENT. THE OFFICIALS HAD LITTLE TO NO UNDERSTANDING OF WHAT THE COAST WAS FOR THE FISHING COMMUNITY.

SMALL-SCALE FISHERS ARE LARGELY LEFT OUT OF POLICY DEBATES, AND OUT OF DRAWINGS. THE DIVERSE RANGE OF SMALL-SCALE PRACTICES ARE CHARACTERIZED AS "TRADITIONAL" OR "ARTISANAL" PRACTICES AND PAINTED AS INEFFICIENT AND TECHNOLOGICALLY UNSOPHISTICATED.

THE NEARSHORE WATERS WHERE THESE COMMUNITIES FISH APPEAR AS A UNIFORM AND BLANK BLUE IN THE COASTAL ZONE PLANS.

A Terranean Coast

The separation of land and sea is an outcome of the CZMP drawing process. Drawing accurate coastal plans has always been a central concern of the CRZ policy. As I followed the surveyors mapping the coast, it was clear that accuracy was much more than just the technological capacity of the survey instruments, or calibrating them to take accurate readings, or reducing instrument error. It was produced in different ways in the process of creating the CZMPs: as the surveyors walked the shoreline, they carefully ensured that visual ambiguities in the satellite image, which serves as the base map or canvas used to trace the CZMP, did not carry over as errors in the final drawing. As the CZMP was made by a team of surveyors who had to operate independently, the team ensured accuracy through recording practices that reduced communication errors.

These ways of drawing, the relations between humans and measuring instruments and practices, and the relation between the CRZ and older surveys have reshaped the coast in important ways. Ultimately, this stress on creating accurate CZMPs meant that the CRZ ended up articulating a *terranean coast*, organized as a set of classified, geo-located elements carefully recorded on the plan. The CZMPs were organized around issues of dividing property, delineating boundaries on land, and relating these to existing land surveys such as city plans, cadastral surveys, and topographic surveys. Thus, they reterritorialized the coast in terms drawn from land governance—as a fixed, divisible, spatial, and terranean entity separated from the sea. Consequently, in the CZMPs, the water is a big swath that, but for some maps that locate corals and sandbars, is largely devoid of detail. These landward views have resulted in large-scale infrastructural interventions that have radically altered the country's coastline and displaced human and nonhuman communities. The complete schism between the land and the sea has greatly affected the human and nonhuman communities who live within the liminal continuities at the shore (Bhattacharyya 2018). It has a drastic effect on nearshore fishing worlds as they mire access to coastal commons in questions of property and land control (Kamath and Dubey 2020). When the 2018 CRZ draft notification was made public, fishers and their regional collectives from across the country organized a protest under the banner of the National Fishworkers Forum (NFF). The fishers pointed out that not only had the government rolled back numerous environmental safeguards, but they had also gone back on the assurances they had made to the community under the older notification. Moreover, the notification was released in English, without many opportunities for public consultation, and it was clear that the draft was designed in a manner that aided the government's infrastructural programs.

In the archives of the Tamil Nadu Coastal Zone Management Authority's website are two very different plans. The first map, approved in 1999, inscribes the coast as a thin set of hand-drawn lines. The base map is traced from a topographic survey sheet, but it is a copy of a copy, specifically, a diazotype or ammonia print, which gives the paper a blue hue.[24] A light mauve wash indicates the Bay of Bengal on the east. A series of lines indicate different limits in the CRZ such as the high and ow tide lines, the two hundred- and the five hundred-meter marks along with existing creeks, roads, and major highways. The different subzones within the CRZ appear as bright yellow and green washes over areas hatched in different textures. Apart from the washes, small corrections in ink and ballpoint pen and marks seem to stand out. The rest of the information are blueprints of pencil lines and letters stenciled by hand. Against the tones of the wash, one can see where the hands faltered, or the stencil was improperly placed.

In sharp contrast, the new map, which is plotted on a sheet of bright white paper, shows the landward side of the coastal zone in much greater detail. A creek, invisible in the old sheet, shows up in bright blue along the coast, as do individual plots that are clearly identified by their numbering. While the new map deletes certain information, such as the presence of coral reefs and estuaries, it connects the CRZ with several other infrastructures and databases. Networks of red lines indicate district roads and railway lines; revenue sources such as saltpans and agricultural plots begin to appear. This difference is most apparent in the index of each map: in the older map the index refers to the topographic survey sheet that the surveyors used to draw the CRZ, shows a key to the state districts, and states that all boundaries are approximate. The index of the new map embeds the CRZ in other survey and enumeration projects. Not only does it link the CRZ to cadastral survey records, it also connects it to more recent satellite surveys, the topographic survey records, recent satellite data on tidal action along the Indian coast, and includes the older CRZ map that preceded it and a key map showing the extent of every village in the district.

Each map functions as a different kind of "graphic artifact" (Hull 2012), which lends the CRZ its bureaucratic authority not just as a result of its content, but also as a result of its material qualities. Unlike the older map, the 2011 map's crisp printed lines add to the kinds of assurances made by government officials regarding the CZMPS—that they, via the capacity of satellite technology, present an accurate picture of the coast. Even as these documents are supposed to render coastal governance transparent and clear up ambiguity, they end up obfuscating and creating new problems. For instance, the state governments took several years to produce the CZMPs after the 2011 CRZ. By the time the CZMPs were approved (some states are still under dispute), the 2018 CRZ had passed, which means that the maps would have to be drawn all over again. The differences in the two maps of Tamil Nadu's coast shows that with each redrawing, the CRZ commits to coast to an extractive and terranean order. If the bluish hues of the old map hint at the possibility of looking toward the aqueous, this possibility is harder to see in the sharp white glare of the newer map.

TIDE LINES AND LIVES

Fishing and Drawing a Salty Shore

ON A SUNDAY MORNING IN JULY 2012, THE RESIDENTS OF MORAGAON KOLIWADA, A FISHING VILLAGE IN THE POSH SUBURB OF JUHU, AWOKE TO THE LOOMING PRESENCE OF THE MV *PAVIT*, A 1,000 TON, 77-METER-LONG OIL TANKER. *PAVIT* WAS ABANDONED BY ITS CREW IN THE GULF OF OMAN AND NOW ITS GHOSTLY PRESENCE HAD SETTLED IN JUHU BEACH.

UPWARD TRENDS IN UNCERTAIN WATERS

THE COASTGUARD JUDGED IT TO BE A SECURITY BREACH AND PROMISED MORE RIGOROUS MARINE SURVEILLANCE PATROLS.

FOR THE CITY'S FISHER COMMUNITY, THIS MEANT HEIGHTENED SURVEILLANCE OF COASTAL WATERS AND THERE WAS THE DANGER THAT *PAVIT* WOULD MOVE WITH THE TIDE AND CRASH INTO THE VILLAGE.

IF IT WAS NOT CLEARED FAST ENOUGH, THE OIL LEAKING FROM THE TANKER WOULD AFFECT SALES RIGHT AT THE BEGINNING OF THE NEW FISHING SEASON.

For communities that depend on nearshore fishing, events such as a beached tanker introduce a sharp sense of instability in a livelihood that is inherently characterized by unpredictability and risk. Unlike intensive fishing practices, nearshore fishers get a far less quantity of catch and operate on slim margins. These events are windows into the fundamental disjuncture between nearshore fishing and contemporary coastal management as institutions and practices that produce coastal worlds. The disjuncture is apparent in the agency they confer to the nonhumans of the shoreline: for instance, while the fisher community saw the *Pavit* as an agent that caused a chain of reactions set within a web of socioeconomic and ecological relationships, the Indian Coast Guard treated it as an agent that had breached a line, crossing into India's territorial waters (Singh 2011).

The idea that the coast is a bounded entity is the direct consequence of policies that organize it as a territorial unit where resources can be mobilized and the borders of which can be secured and controlled. The Coastal Regulatory Zone policy (CRZ), which governs all developmental and conservation activities at the shore, is a vital part of the constellation of political instruments that affect the fishing industry and the human and nonhuman inhabitants of the shore (Menon and Kohli 2008; Menon et al. 2015). The CRZ has a profound influence on the coast because it determines the very definition of what constitutes it, what elements (human, animal, plant, infrastructural) belong to it, and the domain where coastal laws operate. The CRZ's understanding of the coast, as the last chapter shows, is based on the idea that the coast contains multiple elements that can be harnessed as resources for development. Moreover, because the CRZ's focus is land oriented, it reorganizes and appropriates the coast through modes of managing and governing property, remaking it in terranean terms. Coastal Zone Management Plans (CZMPs), cartographic drawings that are central to the CRZ, play an important part in this landward reordering.

An important outcome of this terranean reordering is that it turns the coast into bounded areas that can be disciplined to prevailing policy perspectives. As far as policies that concern Indian fisheries go, the CRZ outlook has always been growth oriented (Bavinck and Johnson 2008; Kurien and Achari 1990). A terranean logic, when imposed on the watery side of the coast, loses conceptual depth. It renders the waters sensible as extractive planes—for example, the continental shelf (the portion of the continental land mass that continues under water, which encompasses 530,000 square kilometers) and the Exclusive Economic Zone (or EEZ, an area of the water up to two hundred nautical miles from the shore over which sovereign states have exclusive rights over marine resources, which in India's case comprises 2.02 million square kilometers) ("Fisheries Profile of India" 2018).[1] These areas become the basis for calculating the potential of the coastal and marine fisheries industry.

The fisheries industry policies are informed by a long-held belief that there is a lot more fish out there than is captured annually.[2] This is a legacy of the first Blue Revolution that began in the 1950s when the government initiated schemes to modernize fishing boats to increase the annual fishing yield (Bavinck and Johnson 2008). This trend has continued in recent policies despite evidence of increasing signs of stress on fish stock, which has hit community-based small-scale fishers the most, pushing them into continuing cycles of debt and poverty (Salagrama 2006). Programs like the second Blue Revolution, which was launched in 2015 by the Department of Fisheries, translate the area of fishing zones into potential catch—a number that is supposed to show yearly growth.[3] This is most apparent in the graphs that are published both at the center and the state level—year after year, production and profit graphs are expected to maintain their upward trend. As these lines follow their prescribed vertical trajectory, they undo the complex human-nonhuman relationships that sustain small-scale nearshore fishing communities and fishing economies.

Unlike intensive fishing practices, small-scale fishing works along temporalities and a mesh of interdependence that entangle the fisher community with fish markets; elemental continuities between mud, marsh, and water; and fragile more-than-human worlds (Todd 2014; Tsing 2015). Nonhumans and their long-cultivated relationships with the fisher community are vital to the nearshore fishing industry, to tenure arrangements and community relationships, and knowledge flows that sustain local economies. This contrasts the anthropocentric, growth-based outlook of the fisheries sector, which has had a drastic effect on coastal ecologies—not just in Mumbai, but in many fisheries across the world as fast depleting fish stocks are making nearshore, small-scale fishing unviable as a livelihood (Green 2020).

At this juncture, it is important to underscore that in contrasting nearshore fishing with extractive regimes of coastal management, I do not wish to replicate problematic dichotomies that characterize indigenous communities and traditional practices as less technological or less sophisticated. I also do not wish to suggest that profit is not a factor in small-scale fishing—it most certainly is. Small-scale fishing has its own issues of asymmetric distribution and cannot by any means be taken as an ideal counter to the state's plans. Instead, in bringing the two together, I position small-scale fishing as a means of getting at other conceptual models of the coast (Subramanian 2009). I take it as a technologically sophisticated and community-oriented enterprise that has much to offer planning and policy. It is also important to note that small-scale and nearshore fishing are not interchangeable terms and they both index diverse practices. Indeed, understanding their specificity is key to developing a better understanding of coastal worlds and human-fish relationships (Jadhav 2018).

Amelia Moore (2019) writes that maritime anthropology has an analytic tradition of linking specificities of nature with cultural specificities, a framework that is ultimately limiting. Instead, Moore articulates the relationship between fishers and fish as multifaceted and encompassing natural, social, and economic systems. Seen this way, fishing is an occupation that is not just characterized by risk due to ecological factors, but also due to the socioeconomic conditions and institutional arrangements in which fishing happens. These multifaceted systems are crucial to coastal nonhumans as they shape the relationship between fishers and the fish. Looking at human-nonhuman relationships as they manifest in nearshore fishing or in the pursuits of coastal management is a way to understand how coasts exist and are constantly worked on as fisheries. *Fisheries* is a term that simultaneously refers to practices of fishing (industrial and recreational fishing, or fishing for subsistence, and the technologies and methods used) and to the area and conditions under which fishing happens, that is, the coast itself. They are, as Moore (2019) suggests, landscapes that change with shifting physical and sociopolitical conditions, and with technologies and practices of fishing. Which is to say, fisheries are not simply an ecological condition—that there are fish out there waiting to be caught—but that the coast and coastal waters and the fish are an outcome of *how* the coast is fished. Fisheries are an outcome of design where even what constitutes "fish" is not a stable term in any sense.[4]

This chapter examines the intersection of two coastal worlds—that of nearshore fishing and of the one crafted by the policies of the fisheries sector. The first part of this chapter explores the coastal fisheries of community-based nearshore fishing. It focuses on nearshore fishing and fish markets and shows how human-nonhuman relationships figure in practices of managing risk in the act of catching fish. These practices do not simply stop with the act of catching fish but continue into the fish markets where the catch must be sold as a product. Setting the right price for the fish involves complex calculi, including projecting the wholesale price at the main market, calculating incurred costs versus profit margins, and a host of other factors such as accounting for external events like beached tankers. The fishing community deploys a range of tactics to engage both economic and ecological risks to turn a profit. Nearshore fishing creates sophisticated fisheries defined by human-nonhuman interdependencies.

When measures such as the CRZ policy are passed, they introduce a new range of vocabularies and concepts that refigure the fishing community in different ways—as stakeholders, wardens of the coast, as knowledgeable subjects, and as vulnerable subjects who stand to lose their livelihoods with coastal change. It also refigures nonhumans—as catch, as bycatch, as coastal buffers against tsunamis, as economic resource, and as vulnerable ecological formations. These reconfigurations introduce new sets of pressures and mitigating conditions into the already dynamic field in which fishers operate and fish spawn.

To legitimize the upward moving graphs that defines the fisheries sector and practices of coastal management, the state must first determine the boundaries of the coast—the area of the zone where it operationalizes fish yield calculations— through coastal surveys. The CRZ policy and its surveys play a vital part in this process. In this drawing process, formations such as berms, dunes, salt, and debris act as signifiers that demarcate new territories in the liminal space where land and sea meet. These nonhuman agents often resist interpretation through the many unexpected ways in which they behave or appear, making it difficult to delineate new boundaries and zones. Consequently, in making the CZMPs, the surveyors must engage in a range of interpretative practices that redistribute nonhuman agency as they measure salinity, delineate forests, and fix tides.

The coast-as-fishery that the CRZ brings forth is radically different from the one that sustains nearshore fishing. It imposes new vocabularies of belonging and value; it reins in coastal waters to legitimize intensive fishing practices, which have long-term environmental consequences. Nearshore fishing offers a strong critique of this growth-oriented outlook—not just in terms of the rights of indigenous fishing communities, but also for the fisheries it creates, which are community oriented and far more sustainable in the long term. It offers possible inroads into decolonizing the knowledge frameworks that undergird coastal industries and conservation and management programs because nearshore fishing offers concepts that are different from the established ones that prefigure ecological relationships at the shore (Green 2020).

Catching Fish

I conducted my fieldwork in Malvani Koliwada, which is in the suburb of Malad in Mumbai, between 2011 and 2012. During this time, I worked with Dev Koli's family, and he operated a "mechanized" boat, which is a term used to describe a range of vessels. Broadly speaking, they are boats with onboard motors that are managed by a small crew; they lie between large boats like trawlers and smaller boats with outboard motors or small crafts that are manually operated.[5] In villages like Malvani, most fishermen operate mechanized boats that are retrofitted with diesel engines of anywhere between 50–160 hp, with a speed of about three to four kilometers per hour.[6] These boats are generally used for day-to-day fishing, and unlike trawlers, which are much larger vessels, lack the capacity to remain at sea for longer durations of time. The Maharashtra government's Fisheries Department describes these small-scale practices as "artisanal" fishing, though there is no exact definition for this term—it refers to a range of fishing techniques used in small-scale fishing.

I accompanied Dev Koli's family while sorting and selling fish at the boat-landing site on the beach. Fishing is an activity divided by gender—it is almost always men who are involved in the process of going out to catch fish while women usually sort and sell fish. In recent years, this pattern has changed, wholesale and export markets have expanded in the city because of which more men have taken on the role of wholesalers, refrigerators, or middlemen and suppliers of fish.[7] While fishing is generally regarded as a traditional occupation of the Koli community, with the expansion of the fisheries industry this is changing as well. Even the small-scale fishing industry employs seasonal migrant workers as deckhands, who return to their villages in the monsoon. The team comprised Dev Koli, his wife, and a crew of three to four men who caught fish, and two to four women who would help sort and sell fish. I would help sort the fish when the catch arrived on the shore, separating different varieties of fish into piles, washing and arranging them in wicker and baskets, and hauling them up from the beach to the road to sell to dealers who would ice the fish overnight. From the icing facility, the fish would be transported to various markets where most of it would be sold wholesale.

Around the time I first began my fieldwork, according to the 2010 Marine Census of Maharashtra (2012), there were 40,953 fishers in the Greater Mumbai district. Of these, only 4,698 identified as active fishers—those who were registered fishers directly engaged in the work of catching fish or collecting fish seed (both full and part-time). Many fishers (20,757 according to the census) worked in allied industries; for example nearly 9,000 women were engaged in fish selling. Most of the boats (4,895 out of a total of 5,725) were mechanized, and of these, 988 used bag nets (also called dol nets), which is the kind of fishing Dev Koli practiced. It is important to note that while these numbers present a general picture of the number of people who use bag nets and the gendered activities in the industry, they may leave out fishers who are not registered with cooperatives, foragers, or seasonal migrant workers.

Mumbai's small-scale bag-net fishers follow a complex system of marine tenure in which each village controls one part of the water where boats from a particular village fish. Many of these small-scale mechanized boat operators use fixed nets (called bag nets) for fishing where a conical net is attached to a buoy that keeps it afloat. The buoy is anchored by stakes located at specific depths in the ocean. Each village controls a fixed share of marine space that is subdivided into set spots that are associated with each boat. Such a fishing process is very different from that associated with itinerant marine vessels, which travel around in the sea to find and catch fish. Larger vessels like trawlers drag large nets in their wake; they dredge the seafloor, bringing up all kinds of fish, including protected species, and destroy breeding grounds and habitats.[8]

Unlike bigger boats that trawl or round up large shoals of fish, bag-net fishers catch fish because of tidal currents. These are like other fixed-net fishing practices such as those that use nets at fixed spots along the mouth of creeks, which are then lowered into the water and trap the fish as they move with the tidal currents. Because bag-net fishing divides the shallow waters and restricts the area where fishing happens, not everyone gets the same kind of catch. This system results in some asymmetry in terms of the species distribution and in pollution load. For example, boats from Malvani made most of their money from catching shrimp, while other villages would catch other kids of fish depending on where their stakes were zoned. The amount of garbage or jellyfish blooms that the Malvani fishermen encountered also varied across seasons and the tide. Gauri Bai, Dev Koli's wife, described this as an asymmetric system in which location, knowledge, and luck were all key factors: some would get more pomfret (which fetched a high price) but then their catch was not guaranteed; other fishers caught different kinds of fish depending on their location. Bag-net fishing, Dev Koli explained, involved so many small and large decisions that one could not pinpoint any one factor and explain the difference in catch between villages or even between different boats. They would both often say that it really depended on that intangible, unknowable substance that is a fisher's luck.

Luck and asymmetries of distribution, as I learned over the course of the season, could be thought of as circumstances beyond human control—as belonging to a realm of outcomes that come about often due to the actions of nonhumans such as the tide and the fish. To harness this luck, fishers had to cultivate a close relationship with the water by understanding its currents through its colors and behavior over the year; they also had to learn the movements of the fish and accumulate knowledge of features such as sandbars and hills and trees visible from the shore. In such community-based fishing practices, success depends on cultivating relationships and harnessing place-based knowledge of the coast (García-Quijano 2007).

While bag-net fishing divides the coast into distinct territories, the fishing area allotted to individual villages is further divided based on depth. At set points in the sea, the fisher community installs stakes into the muddy seafloor, which are marked by buoys. Depending on the time of the month and tidal strength, fishers make decisions about the depth at which to fish. Dev Koli's boat, *Rupa,* had stakes set at roughly five, seven, and ten *wav* (a unit of measure described as a fathom). He, like many others, had recorded the location of these stakes with a GPS device, though his pilot could navigate to it without its aid. If the current was very strong and the waves were high, Dev Koli would fish at a lesser depth. If it seemed like the creek and the sea were churning less, they might venture further out. Fishers like Dev Koli were keen readers of the water. His pilot Gopal was especially adept at reckoning conditions from the color of the water and the tactile feedback he would get from navigating the boat. He also knew the features of the creek intimately, navigating in and out with practiced ease.

While the description of marine tenure arrangements along Mumbai's coast might appear as a stable and preset arrangement as it completely does away with the problem of finding fish, this does not eliminate all the unknown or unstable elements that make up fishing. As no one boat or fisher working alone could possibly hope to be successful, in such a scenario, "active knowledge-making" (Hoeppe 2007, 43) is vital to a fisher community's success. This knowledge-making involves gathering information on an individual level (such as checking the height of the tide and keeping a track of the weather) and at the level of the collective, which takes the form of fast snippets of information that pass between boats. It relies on assessing events and conditions as they come into being, and on crosscutting and analyzing that information in relation to past experiences of similar events, seasonal occurrences, and cycles. Understanding how small-scale fishing works is instructive because it offers a way of attending to "fish pluralities," a term Zoe Todd (2014) uses to refer to the many ways of knowing and thinking about fish (and in this case, by extension, the coast). In this sense, bag-net fishing is not just a community-oriented practice, but also a coast-oriented practice that reaches across species divides, a process that is grounded in the material qualities and temporalities of tidal, geological, and seasonal systems.

Every morning, Dev Koli would get text messages that would let him know which boats had decided to set out or were about to set out. He would confer with his fellow fishers at the beach regarding the depth, but like he said, it was not always necessary to do what everyone else was doing. Since bag nets are nets that are fixed to a static point and not dragged behind a boat, fishers rely a lot on the force of the tidal current to keep their nets afloat.[9] With respect to tides, fishers have to be wary and attentive both in terms of the daily rhythms and the monthly lunar cycle. India's west coast experiences mixed semi-diurnal tides, which means that it gets two sets of high and low tides of varying magnitude every day. Though, theoretically, this provides fishermen with two sets of opportunities on any given day to cast their nets, nearly all bag-net fishing happens during the stronger set.[10]

Apart from the daily highs and lows, the tidal forces follow a cycle based on the lunar month where they move between *jowar* (stronger spring tides) and *bhang* (weaker neap tides). As the force of the tidal current is very low during neap tides, it is a challenge to keep the nets open, and the waters tend to be more still. At this time, the fish are thought to be quieter as the water does not churn them and move them into the nets. During this period, fishers have to account for the cost of going out to fish against a future potential catch. Thus, fishers must make quick decisions about whether to cast their nets, especially in the days around the first and third quarters of the moon based on weather conditions and by paying attention to the decisions and actions of others.

Bag-net fishing weaves particular kinds of interdependencies into the fisher community. For example, installing fishing stakes, the basic infrastructure for bag-net fishing, is a collective enterprise where at least two boats have to come together to drive the stake into the mud. A stake is attached to a long wooden pole (usually a palm tree log), which is suspended between two boats that anchor themselves into position. The stake is then driven into the ground using the undulating force of the waves that moves the boats up and down.[11] Often times, if a fisher was unable to take the boat out for any reason, others would help them out by casting a net at their buoy and hauling in their lobster traps.

These responsibilities, obligations, and care are not just extended within the community of fishers, but to the coast itself—managing asymmetries of distribution, ensuring that there is enough fish left for others, ensuring that fish are left to spawn for coming years, and caring for the creek and creek life. As state policies and fishing technologies change, they introduce new conflicts that affect human-fish-coast relationships: for example, during my fieldwork, bag-net fishers would often complain to municipal officials that fishers who could afford to buy larger, more powerful boats were depleting fish stocks at a faster rate than fishers with older, slower boats. They would urge officials to put caps on engine capacities and on intensive fishing to maintain coastal equity.

The knowledge building and improvisation that undergirds bag-net fishing happens on multiple levels and are not just confined to the act of catching fish, but also to selling the catch: not only do the fishermen and women have to exchange information about tides, currents, and the day-to-day behavior of fish, they also have to keep a constant look out for the catch hauled in by various boats and project the potential price tag they can hope to pin on their catch. In addition to engaging in this process of gathering and disseminating information, fishers also must make several crucial economic decisions simultaneously. Knowledge gathering, projecting, and cultivating ties are important aspects that govern fishing as well as selling fish.

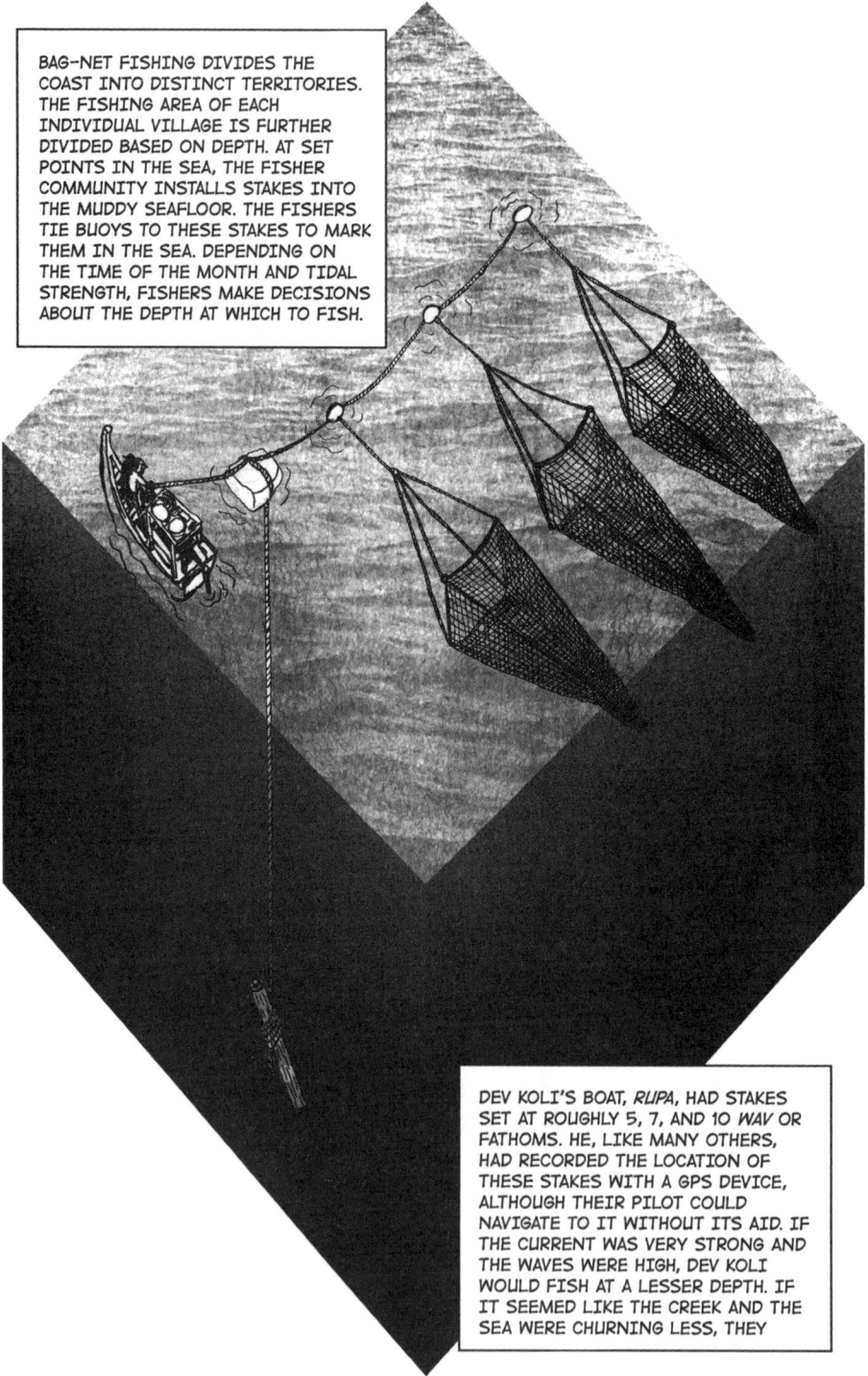

BAG-NET FISHING DIVIDES THE COAST INTO DISTINCT TERRITORIES. THE FISHING AREA OF EACH INDIVIDUAL VILLAGE IS FURTHER DIVIDED BASED ON DEPTH. AT SET POINTS IN THE SEA, THE FISHER COMMUNITY INSTALLS STAKES INTO THE MUDDY SEAFLOOR. THE FISHERS TIE BUOYS TO THESE STAKES TO MARK THEM IN THE SEA. DEPENDING ON THE TIME OF THE MONTH AND TIDAL STRENGTH, FISHERS MAKE DECISIONS ABOUT THE DEPTH AT WHICH TO FISH.

DEV KOLI'S BOAT, *RUPA*, HAD STAKES SET AT ROUGHLY 5, 7, AND 10 *WAV* OR FATHOMS. HE, LIKE MANY OTHERS, HAD RECORDED THE LOCATION OF THESE STAKES WITH A GPS DEVICE, ALTHOUGH THEIR PILOT COULD NAVIGATE TO IT WITHOUT ITS AID. IF THE CURRENT WAS VERY STRONG AND THE WAVES WERE HIGH, DEV KOLI WOULD FISH AT A LESSER DEPTH. IF IT SEEMED LIKE THE CREEK AND THE SEA WERE CHURNING LESS, THEY

AFTER SOME MONTHS OF SORTING FISH AT THE BEACH, I FINALLY GOT A CHANCE TO SEE HOW BAG-NET FISHING WORKED.

I MET MAMA AND HIS CREW EARLY IN THE MORNING. THE TIDE WAS ALREADY RECEDING AS WE WADED INTO THE WATER TO REACH THE BOAT. MAMA HAD TALKED TO A FEW OTHERS WHO WERE PLANNING TO VENTURE OUT THAT DAY AND LIKE THE OTHERS, HE HAD DECIDED TO TRY HIS LUCK AT A DEPTH OF 7 *WAVS*. NOT MANY BOATS WERE HEADING OUT AS WE WERE FAST HEADING INTO THE *BHANG* PERIOD WHEN THE CURRENTS WOULD BE AT THEIR WEAKEST.

I SAT ON THE RIM OF THE BOAT AND WATCHED GOPAL, THE PILOT, AS HE GUIDED THE VESSEL OUT OF THE CREEK AND INTO THE OPEN SEA. HE DEFTLY MANEUVERED THE BOAT TO AVOID SANDBARS AND ROCKS THAT LURKED BENEATH THE MUDDY WATERS. GOPAL WAS REGARDED AS AN EXPERT PILOT IN THE VILLAGE, WITH AN INTIMATE KNOWLEDGE OF THE INS AND OUTS OF THE CREEK, AND HE WAS AWARE OF EVERY LANDMARK AND FEATURE OF THE DYNAMIC WATERWAY.

DEKH, AAJ APAN KO LATE HO GAYA... [LOOK, WE ARE LATE TODAY . . .]

WHILE OTHERS WOULD CLAIM THAT HE COULD, AND HAD ON SEVERAL OCCASIONS NAVIGATED THE BOAT UNAIDED ON THE DARKEST OF NIGHTS, GOPAL HIMSELF WOULD BRUSH OFF THESE STATEMENTS AND ATTRIBUTE SUCH EVENTS TO LUCK.

NAMASKAR! TUMHI NIGHUN GELAT KA? HO, AMHALA USHIR ZALA AHE . . .

[HELLO! HAVE YOU ALL LEFT? YES, WE ARE LATE . . .]

ONCE, WHILE SORTING FISH, I HAPPENED TO ASK HIM WHAT ONE HAD TO DO TO GAIN SUCH LUCK AS HE SEEMED TO POSSESS.

TO THIS, HE HAD REPLIED THAT LUCK WAS THE FINAL, MOST TRIVIAL INGREDIENT OF A PROCESS THAT INVOLVED AMASSING EXPERIENCE OVER TIME.

GOPAL EXPLAINED THAT ONE LEARNED THE POSITION OF SANDBARS AND ROCKS UNTIL ONE KNEW THEM INTIMATELY ENOUGH.

ONE ALSO LEARNED TO UNDERSTAND THE BOAT AND THE WATER BY MUSCLE AND TOUCH, FEELING ONE'S WAY THROUGH THE WATERS, KEEPING AN EYE OUT FOR OBSTACLES, AND RESPONDING TO ANY SUSPICIOUS RESISTANCE FROM THE BOAT'S MECHANICAL STEERING ROD.

CASTING HIS EYES OUT AT THE WATER, HE SAID HE KNEW WHAT THEY (THE WATER AND BOAT) WOULD BE LIKE AT DIFFERENT TIMES OF THE MONTH AND OVER THE SEASONS.

ONBOARD *RUPA*, DEV KOLI PULLED OUT HIS GPS DEVICE AND SHOWED ME THE BOUY'S LOCATION. AS GOPAL GOT US OUT OF THE CREEK AND INTO THE SEA, DEV MAMA SHOWED ME HOW THE SCREEN PROVIDED DIRECTIONS. HE SAID THAT THE SEA WAS LIKE AN OPEN ROAD, BUT WITHOUT ANY REFERENCE POINTS. THUS, THE GPS WAS MOST HELPFUL IF THEY WERE GOING OFF-COURSE, ESPECIALLY IN TIMES OF POOR VISIBILITY.

MAMA SAT ON A RAISED PLATFORM ABOVE THE ENGINE AND CYCLED THROUGH THE SAVED LOCATIONS ON THE DEVICE, MANY OF WHICH WERE OF BUOYS AND LOBSTER TRAPS THAT BELONGED TO OTHER BOATS. IF ANY OF MAMA'S FRIENDS NEEDED HIS HELP OR COULDN'T GO OUT ON A PARTICULAR DAY, HE COULD PROVIDE THE NECESSARY ASSISTANCE. SIMILARLY, OTHERS HAD MAMA'S BUOYS AND TRAPS STORED ON THEIR GPS DEVICES.

VINOD, ONE OF THE DAILY WAGE LABORERS, SAT DOWN AND REMOVED ONE OF THE MANY WOODEN PLANKS THAT FORMED THE BOAT'S DECK. HE PULLED OUT A LITTLE STOVE AND MADE TEA FOR THE CREW.

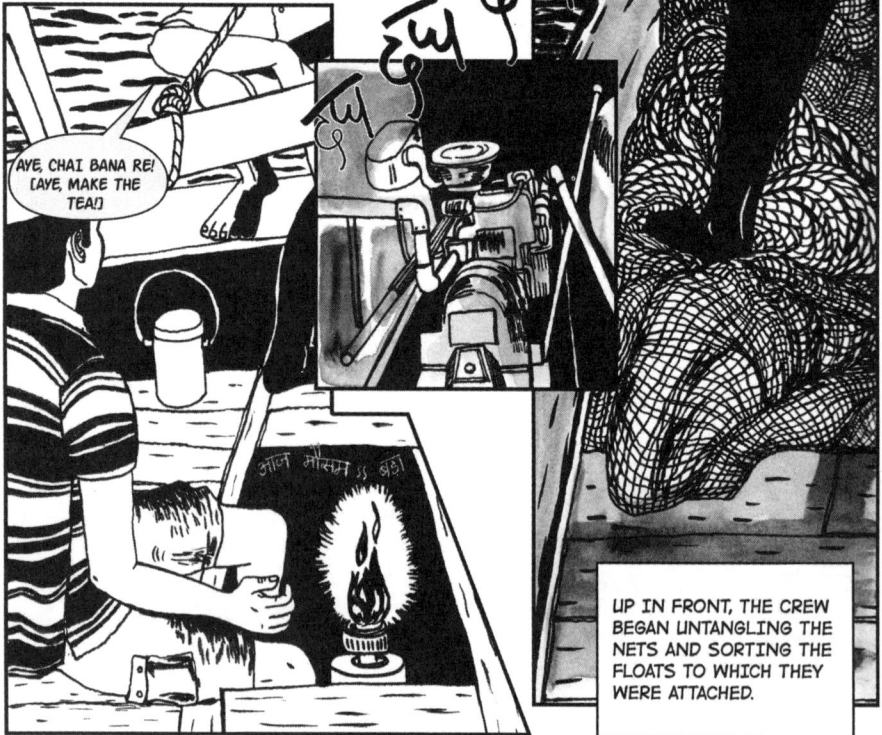

AYE, CHAI BANA RE!
[AYE, MAKE THE TEA!]

UP IN FRONT, THE CREW BEGAN UNTANGLING THE NETS AND SORTING THE FLOATS TO WHICH THEY WERE ATTACHED.

THE WATER, WHICH WAS A MUDDY BROWN AT THE SHORE, WAS NOW BLUE AND CLEAR. THE BOAT BOBBED UP AND DOWN NEAR THE BUOY AND VINOD REACHED OUT AND DRAGGED THE ROPE ATTACHED TO IT. THE MEN PULLED OUT THE LONG NETS FROM THE FRONT OF THE BOAT AND LINKED IT TO THE ROPE. THE CURRENT CARRIED THE NETS AWAY FROM THE BOAT AND HELPED THEM UNFURL IN THE WATER, WHERE THEY HUNG, GENTLY MOVING WITH THE TIDE.

ONCE ALL THE NETS WERE SET, THE MEN NAPPED FOR A COUPLE OF HOURS TILL IT WAS TIME TO HAUL IT ALL BACK IN AND MAKE OUR WAY TO THE BEACH. IN THE DISTANCE, I COULD SEE A LINE OF OTHER BOATS WHERE CREWS WERE PROBABLY DOING THE SAME THING. I ALSO SETTLED DOWN FOR A NAP; THE VIBRATION FROM THE REPURPOSED TRUCK ENGINE, THE DIESEL FUMES, AND THE SOUND AND ACTION OF THE WATER MADE IT HARD TO STAY AWAKE.

THEY BEGAN HAULING THE NETS BACK IN AFTER THEY WOKE. THE MEN
WORKED QUIETLY, ALL THEIR ENERGY FOCUSED ON THE ARDUOUS TASK OF
GETTING THE NETS BACK ON BOARD. WITH THEIR LEGS PLANTED ON THE
SLICK PLANKS, THEY LEANED BACK TO MAKE THE MOST OF THEIR BODY
WEIGHT. THE ROPES WERE ROUGH AND FRAYED, AND THE SUN WAS BLINDINGLY
BRIGHT AS IT REFLECTED OFF THE WATER.

DESPITE ALL THEIR EFFORTS, THE CATCH WAS POOR. MAMA TOOK OUT A
LARGE JELLYFISH FROM THE NET AND SET IT ON A TUB. HE THEN TOOK OUT A
HALF-EATEN STINGRAY AND SET THAT ASIDE AS WELL. ONCE ALL THE NETS
WERE IN, THE BOAT TURNED BACK TOWARD THE SHORE.

Selling Fish

Women do most of the sorting and selling activity at the beach. I would begin my day by joining Gauri Bai (Dev Koli's wife), who I called Mami, while she sat with a group of other fisherwomen who would gather in the shade of the trucks as the tide turned. As the boats started to come in, they would get a quick sense of the day's haul—not just in terms of their boat, but also in relation to all other boats—visually estimating it to judge potential prices they could set. The beach is a site of continuous assessment, which begins even before the boats return to the landing site with their catch. One of the most important tasks that happened at this time was to square away the previous day's accounts with the wholesale buyer. The women would flock around the few buyers who came to the beach and occasionally demand to renegotiate the price that was agreed on the previous day, haggling when they felt the wholesale dealer had made a steep profit. It was not uncommon to hear complaints of short changing.

While we sat in the shade cast by the trucks, Gauri and the other women exchanged friendly banter that was also an exercise in gaining information regarding the previous day's catch, the negotiated price, and news from the wholesale market, including the going rate for different varieties of fish. Selling involves the active accumulation and management of knowledge; it involves paying attention when others conduct sales and appraising their catch, cultivating relationships with vendors and extracting information about the wholesale market from them, and paying attention to the casual talk that circulates on the beach.

Once the boats arrived, Mami and I would make our way to the water's edge where we would be joined by three to four other women: Indu, who was Gopal-the-pilot's wife; Lata Bai, who was married to Ganesh, a crewmember on the boat; and two daily-wage laborers, Mary Bai and Jenny Bai. The number varied—not all women showed up every day and the daily-wage workers were often hired based on how much catch needed to be sorted and carried. The women would often bring along their children who would sit and sort with us.

Once the boats landed, everything would happen at a great speed: men removed the fish from the nets into tubs and dunked them into the creek to give the catch a quick wash. Some of the catch, which remained in the nets (such as the shrimp) would be washed in the creek as well. Everything apart from the shrimp, which is caught in a separate part of the net, was dumped in one pile on the beach.[12] We would squat around this pile and begin the work of sorting the catch. Inedible fish, like eels or half-eaten fish, were discarded. Once sorted, the separated catch was then arranged in tubs. These tubs were of standard sizes, which made it easy for setting the price. The crew had to move fast to beat the tide rising close to our backsides.

While there are several women who make their living by selling fish in the retail market, or going door to door, most of the women who dealt with the catch at the beach would sell it directly at the landing site to avoid additional costs or losses. The competition was always intense as all the women wanted the best price and to rid themselves of their catch as soon as possible.[13] The main reason for this was that if the fishers were stuck with the catch, they would then have to pay money to have it transported or put the catch in storage. On the few occasions when the dealers seemed to offer too low a price, or if our boat had returned too late to make a sale, all the larger fish would be sent by a truck to be sold in the local market at the nearest railway station. In the event the shrimp did not sell, it would be carted to drying areas north of the beach where Mary and/or Jenny would spread out the shrimp to dry. We would then have to spend the whole afternoon shooing crows while waiting for another agent who dealt solely in dried shrimp. This involved not only the effort of haggling with another agent, but it also meant time lost from housework, paying the workers extra, and the effort of spreading and watching over the catch.

THERE WERE A FEW WHOLESALE DEALERS
WHO WERE REGULARS AT THE BEACH AND
HAD THEIR OWN SYSTEM OF FOLLOWING
THE MARKET AND SETTING THE PRICE AT
WHICH THEY BOUGHT THE FISH.

ONE OF THEM, BINDIYA, ALLOWED ME TO
SHADOW HER TO UNDERSTAND THE
PRICE-SETTING PROCESS. FOR A WEEK, I
SHADOWED BINDIYA AS SHE WENT FROM BOAT
TO BOAT ON HER NEGOTIATIONS.

IF SHE ARRIVED EARLY, BINDIYA WOULD JOIN THE WOMEN SITTING BY THE TRUCKS.
SHE WOULD LISTEN TO THEIR CONVERSATION AND TALK TO OTHER WHOLESALE
DEALERS ABOUT THE MARKET. IF A SELLER WAS PRESENT, SHE WOULD SETTLE THE
ACCOUNTS FROM THE PREVIOUS DAY.

AY! WATCH OUT, LOOK WHO'S HERE! HAHA!

LET HER COME!

HEHEHE...

SHE HAS TO BUY ALL THIS FISH FOR HER LUNCH!

Arranging the catch in standard size containers makes it easy to eyeball quantities and set the price

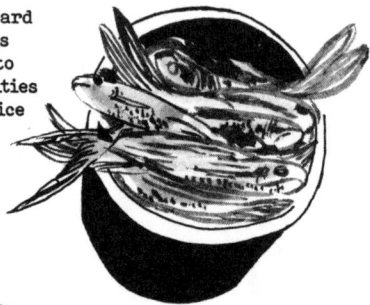

THE SYSTEM OF SELLING THE CATCH TO THE WHOLESALE DEALER WORKED ON THE BASIS OF SHORT-TERM SPECULATION WHERE BOTH THE BUYER AND THE SELLER ENTER INTO AN AGREEMENT BASED ON A PROJECTED PRICE FOR THE CATCH AT THE WHOLESALE MARKET.

WHEN BUYING THE CATCH, BINDIYA WOULD ASSESS ITS VALUE TO ESTIMATE THE PRICE SHE WOULD GET AT THE WHOLESALE MARKET.

LOOK HOW GOOD THIS SQUID IS!

SHE WOULD THEN PAY THE SELLER A LOWER AMOUNT AND MAKE HER PROFIT FROM THE DIFFERENCE. WHEN SHE CALCULATED THIS VALUE, BINDIYA WOULD ALSO TAKE INTO ACCOUNT THE COST OF STORING AND TRANSPORTING THE CATCH, ALL OF WHICH INCURRED INFRASTRUCTURE AND LABOR CHARGES.

NO NO NO IT HAS TO BE MORE THAN THAT!

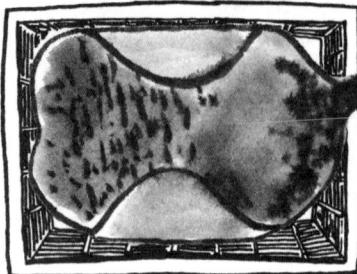

ARRE RE RE!

CHAL 400 DE. [OKAY, GIVE RS. 400]

NAHI RE...KASA...? [NO RE...HOW... (NOT POSSIBLE)]?

BAGH, LIHITES KAI? AYE, USNE KYA LIKHA JAKE DEKH! [GO CHECK WHAT (PRICE) SHE WROTE!]

PARVA TU MALA 400 DEELE! [YOU GAVE ME 400 DAY BEFORE YESTERDAY!]

MI 350 LIHILE! AYE! [I SAID 350! HEY!]

MI TECH DEELE! [I GAVE THE SAME (PRICE)!]

TULA KITI DEELE? [WHAT PRICE DID SHE SETTLE WITH YOU?]

AYE! TEELA KA 400 DEELES? [WHY DID YOU GIVE HER 400 (AND NOT ME)?]

NAI GA.. [NO (IT'S NOT LIKE THAT)]

EK NUMBER LIKHNE KA, DOOSRA NUMBER DENE KA! [SHE WRITES ONE NUMBER AND GIVES ANOTHER NUMBER!]

APART FROM THESE FACTORS, A CRUCIAL DECIDING ASPECT WAS BINDIYA'S LONG-TERM RELATIONSHIP WITH THE FISHERS. THE PROCESS OF FIXING A PRICE WAS ONE THAT INVOLVED A GREAT DEGREE OF HAGGLING, SHOUTING, EMOTIONAL BLACKMAIL, SWEARING, AND CAJOLING. IT HELPED TO HAVE A GOOD RELATIONSHIP AS IT MADE THE BACK AND FORTH A LOT EASIER. IT ALSO BOUND THE FISHERS AND WHOLESALERS IN A MESH OF OBLIGATIONS THAT COULD BE INVOKED IN TIMES OF NEED.

BINDIYA WOULD OFTEN SAY THAT GIVEN THE MARGINS SHE OPERATED ON, THERE WAS VERY LITTLE ROOM TO NEGOTIATE IF SHE HAD TO MAKE A PROFIT.

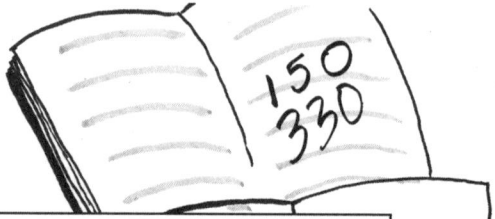

150
330

BUT SHE WOULD ALSO REMARK, WHEN NOT SITTING NEAR THE CROWD, THAT ONCE IN A WHILE SHE WOULD ACCEPT EARNING LESS PROFIT IF IT MEANT ENSURING THAT SHE WOULD HAVE A STEADY SUPPLY OF THE CATCH FROM A BOAT, IN ORDER TO NOT LOSE THEM TO ANOTHER WHOLESALE DEALER – ESPECIALLY IF SHE KNEW THERE WERE MITIGATING CIRCUMSTANCES

560
440

945
700
540

DEKH . . .
[LOOK . . .]

APNA BHI TIME KHARAB JA SAKTA HAI . . .
[ANYONE CAN FALL ON HARD TIMES . . .]

680
845

360
280
175
500
300
180
800
150

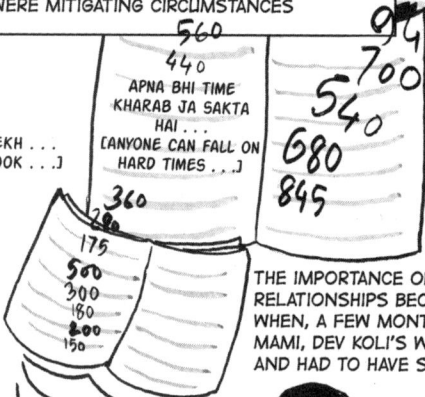

THE IMPORTANCE OF LONG-TERM RELATIONSHIPS BECAME APPARENT WHEN, A FEW MONTHS INTO 2012, MAMI, DEV KOLI'S WIFE, FELL ILL AND HAD TO HAVE SURGERY.

SINCE SHE HAD CONDUCTED ALL THE NEGOTIATIONS WITH DEALERS LIKE BINDIYA, THE OTHERS HAD TO MANAGE IN HER ABSENCE. DEV KOLI WOULD STEP IN FROM TIME TO TIME, BUT IT WAS HARD TO NEGOTIATE WITH THE SAME INTENSITY, PARTICULARLY DURING TIMES OF FIERCE COMPETITION.

PAKAT HAVE KA?
[DO YOU WANT THIS STINGRAY]?

As Jane Guyer (2016) suggests, the price of a commodity is a "composite" that represents "the results of narratives of creation, addition, and subtraction." In the case of fish, one can think of price as a composite of additions and subtractions based on a mixture of factors that operate in the long and short term: general trends in seasonal prices, projections based on daily catch at different landing centers, labor and storage costs, and the protracted history of favors exchanged between different parties. A determining factor in this composite is the material nature of the catch: fish, as a commodity, does not keep unless frozen. The outer shell of creatures such as shrimp and crabs cut into the skin, jellies burn, and under pressure from weather, market, and tide, time is of essence. Especially in warmer days, it was most important to get rid of the fish as soon as possible—there were many days when Mami would just about manage to break even or even take a loss, as the alternative would be to spend more money out-of-pocket to move, store, and find another buyer for the fish.

Being on the beach and cleaning, separating, arranging, and selling fish is physically arduous, and involves learning several bodily techniques to perform these tasks with speed and efficiency. Fish work entails improvisation because markets can dip or rally, fish can move, and they can also rot and stink. Thus, fishy improvisations happen across different scales and registers; for instance, not only do outcomes depend on the action of the team members of one boat, but they also look to the actions of others to facilitate positive outcomes. Thus, actions and decisions themselves cannot be thought of as simply a response to a given set of information, but as occurring in relation to different horizons of aspirations and the desire to secure the biggest catch. Fishing, especially bag-net fishing, is a precarious occupation for many reasons. It is an occupation that can only be carried out at certain times of the year or month, depending on the tide and weather conditions, thus greatly diminishing the chances of having a higher profit margin. Fishers must factor in the rising cost of diesel and repair against this slim margin. Fishers also work in increasingly polluted waters and in harsh conditions. Thus improvisation, decision making, and knowledge gathering are key to not just securing an income, but are also key strategies for absorbing political, environmental, and market volatilities.

It was through conversations on the beach during times of work and waiting that I learned about how the everyday practice of catching, sorting, and selling fish occurs: it involves a continuous process of improvisation based on knowledge acquired over time and based on the immediate context. Within the anthropology of fishing, different authors pose improvisation as a crucial act for maximizing one's catch or profit. For instance, in Alice Volkman's (1994) work on women in fishing communities in Indonesia, improvisation is key to ensuring economic success. Depending on the catch, individual sellers come up with different strategies of processing and selling fish to maximize profit. Improvisation, strategizing, and decision making are critically linked to expertise and mastery—they are also aspects of the social fabric that bind fisher communities, where decisions and actions are not carried out in isolation, but as a collective.

Apart from thinking of bag-net fishing as a means for sustenance, it can also be thought of as a mode in which the coast is made. Small-scale fishing is a process of creating fisheries through long-cultivated relationships within the community (helping each other fish, sharing information, and creating fishing grounds), and with nonhumans (the boats, the coast, and the many different animals that constitute the term *fish*). These relationships and the wells of knowledge they contain are crucial for not just sustaining livelihoods, but also the forms of life and the very materials that make the shore. It harnesses, as Ingold (2010, 10) writes of entanglements, the emergent possibilities in the present, "undertaking a forward movement, which gives rise to things . . . and follows the ways of the world as they unfold." Given that much of this open-ended improvisation is oriented toward maintaining fish stocks, occurs in co-relation to the creek, and draws attention to the asymmetries of human-fish relationships and opportunity distribution, practices such as bag-net fishing stand as interesting conceptual counterpoints to the territorializing logics of the CRZ and provide openings for rethinking resource-based understandings of the coast.

FISHING DECISIONS

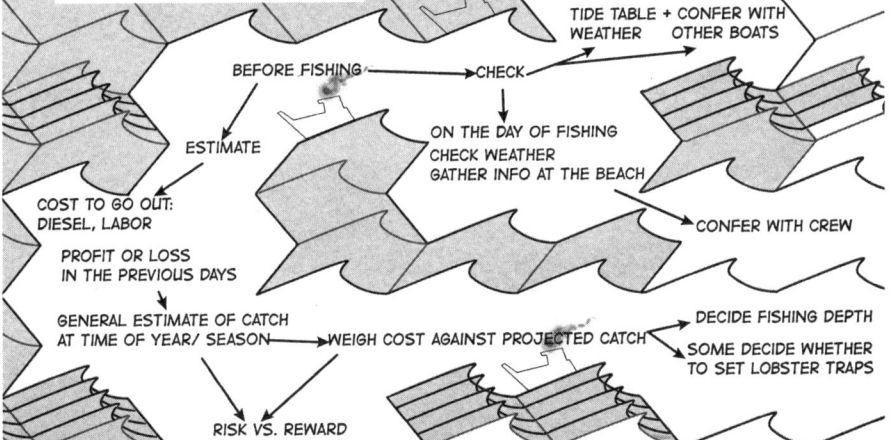

BEFORE FISHING → CHECK → TIDE TABLE + CONFER WITH WEATHER   OTHER BOATS

ON THE DAY OF FISHING
CHECK WEATHER
GATHER INFO AT THE BEACH

CONFER WITH CREW

ESTIMATE

COST TO GO OUT:
DIESEL, LABOR

PROFIT OR LOSS
IN THE PREVIOUS DAYS

GENERAL ESTIMATE OF CATCH
AT TIME OF YEAR/ SEASON → WEIGH COST AGAINST PROJECTED CATCH → DECIDE FISHING DEPTH

SOME DECIDE WHETHER
TO SET LOBSTER TRAPS

RISK VS. REWARD

SELLING DECISIONS

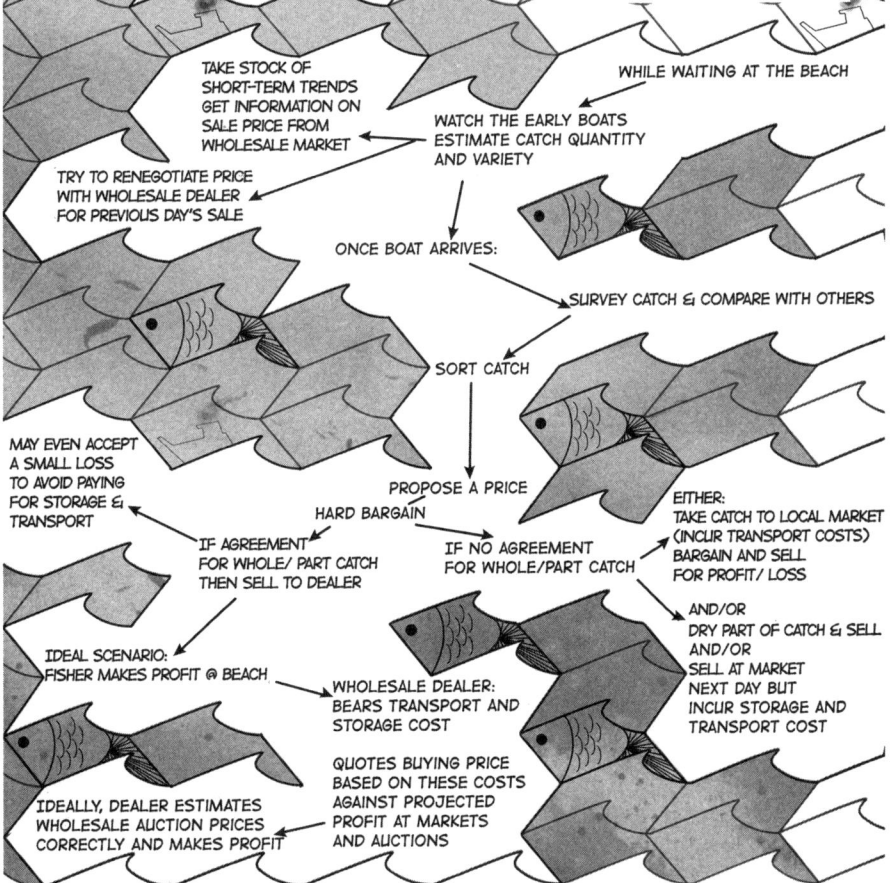

WHILE WAITING AT THE BEACH

TAKE STOCK OF
SHORT-TERM TRENDS
GET INFORMATION ON
SALE PRICE FROM
WHOLESALE MARKET

WATCH THE EARLY BOATS
ESTIMATE CATCH QUANTITY
AND VARIETY

TRY TO RENEGOTIATE PRICE
WITH WHOLESALE DEALER
FOR PREVIOUS DAY'S SALE

ONCE BOAT ARRIVES:

SURVEY CATCH & COMPARE WITH OTHERS

SORT CATCH

MAY EVEN ACCEPT
A SMALL LOSS
TO AVOID PAYING
FOR STORAGE &
TRANSPORT

PROPOSE A PRICE
HARD BARGAIN

EITHER:
TAKE CATCH TO LOCAL MARKET
(INCUR TRANSPORT COSTS)
BARGAIN AND SELL
FOR PROFIT/ LOSS

IF AGREEMENT
FOR WHOLE/ PART CATCH
THEN SELL TO DEALER

IF NO AGREEMENT
FOR WHOLE/PART CATCH

AND/OR
DRY PART OF CATCH & SELL
AND/OR
SELL AT MARKET
NEXT DAY BUT
INCUR STORAGE AND
TRANSPORT COST

IDEAL SCENARIO:
FISHER MAKES PROFIT @ BEACH

WHOLESALE DEALER:
BEARS TRANSPORT AND
STORAGE COST

IDEALLY, DEALER ESTIMATES
WHOLESALE AUCTION PRICES
CORRECTLY AND MAKES PROFIT

QUOTES BUYING PRICE
BASED ON THESE COSTS
AGAINST PROJECTED
PROFIT AT MARKETS
AND AUCTIONS

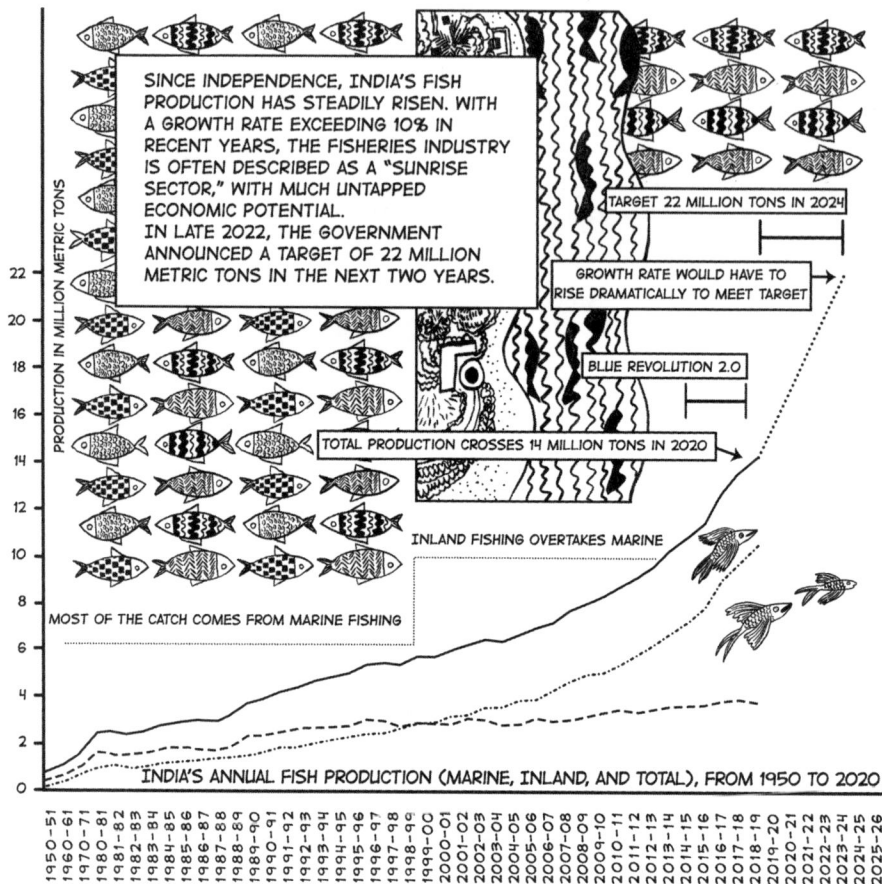

SINCE INDEPENDENCE, INDIA'S FISH PRODUCTION HAS STEADILY RISEN. WITH A GROWTH RATE EXCEEDING 10% IN RECENT YEARS, THE FISHERIES INDUSTRY IS OFTEN DESCRIBED AS A "SUNRISE SECTOR," WITH MUCH UNTAPPED ECONOMIC POTENTIAL.
IN LATE 2022, THE GOVERNMENT ANNOUNCED A TARGET OF 22 MILLION METRIC TONS IN THE NEXT TWO YEARS.

TARGET 22 MILLION TONS IN 2024

GROWTH RATE WOULD HAVE TO RISE DRAMATICALLY TO MEET TARGET

BLUE REVOLUTION 2.0

TOTAL PRODUCTION CROSSES 14 MILLION TONS IN 2020

INLAND FISHING OVERTAKES MARINE

MOST OF THE CATCH COMES FROM MARINE FISHING

PRODUCTION IN MILLION METRIC TONS

INDIA'S ANNUAL FISH PRODUCTION (MARINE, INLAND, AND TOTAL), FROM 1950 TO 2020

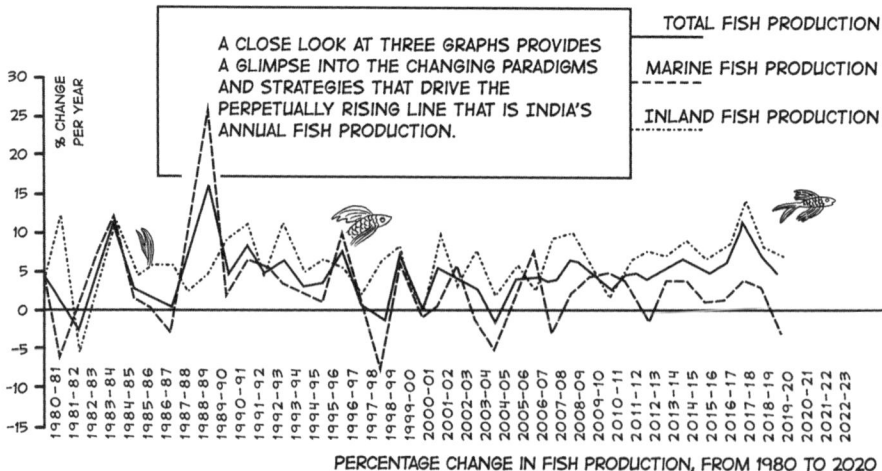

TOTAL FISH PRODUCTION

MARINE FISH PRODUCTION

INLAND FISH PRODUCTION

A CLOSE LOOK AT THREE GRAPHS PROVIDES A GLIMPSE INTO THE CHANGING PARADIGMS AND STRATEGIES THAT DRIVE THE PERPETUALLY RISING LINE THAT IS INDIA'S ANNUAL FISH PRODUCTION.

% CHANGE PER YEAR

PERCENTAGE CHANGE IN FISH PRODUCTION, FROM 1980 TO 2020

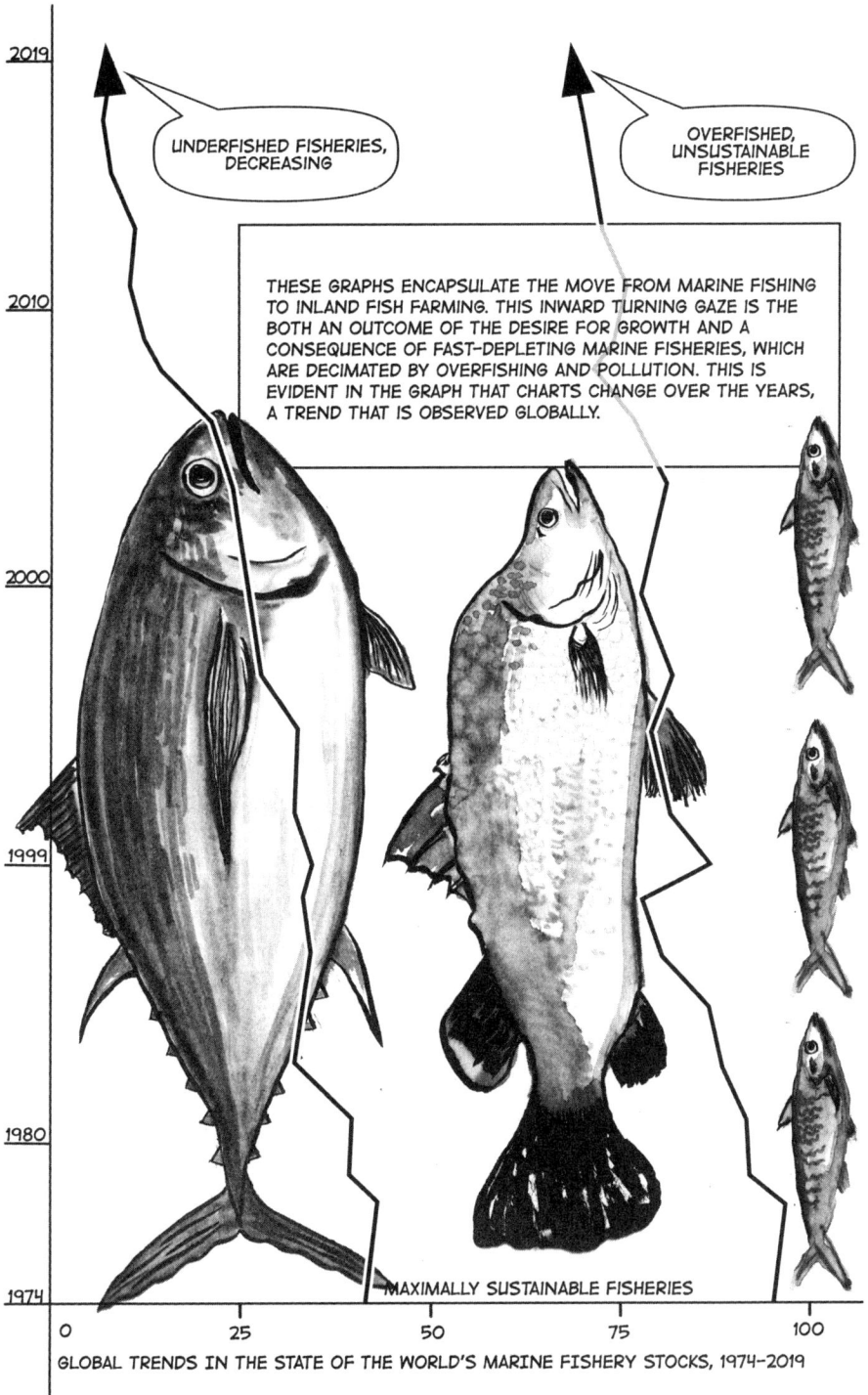

GLOBAL TRENDS IN THE STATE OF THE WORLD'S MARINE FISHERY STOCKS, 1974–2019

Securing the Seas

The zone where the stakes and buoys are attached for bag-net fishing has not shifted much in the last few decades, though these do not figure in any coastal plan published by the state government. Over the years, boats, which used to rely on wind sails, slowly converted to using diesel motors. From the 1950s onward, the Indian state began offering subsidies to fishers to mechanize their boats, which would allow them to move toward higher-yield practices, including large-scale trawler and purse-seine net fishing. The goal of the first Blue Revolution, which was launched at this time, was to industrialize capture-fisheries, and because of these subsidies and technological interventions, the marine fisheries sector grew at a rapid rate (Nair 2021).[14] From the 1990s onward, the government began promoting shrimp farming and consequently, the annual production rates from inland fishing began catching up with marine sources. Throughout this time, the government continued its support of intensive practices such as purssein netting and trawler fishing, despite growing concerns of depleting fish stock.

From 1963 to 2003, the overall fish production in India jumped fivefold, from 0.5 million tons to over 2.5 million tons. Between 2003 and 2015, this number crossed 10 million tons ("Fisheries Profile of India" 2018). With the introduction of the second Blue Revolution, a program that ran from 2015 to 2021, the government invested a great deal of capital in further developing inland fisheries. However, these intensive methods have had a devastating effect on coastal ecology (Pitchon 2015; Wiber et al. 2012). Purse-seine netting and trawling catch juvenile fish and result in a large amount of bycatch—creatures that are deemed inedible but are nevertheless caught in the nets and die. Unlike trawler or purse-seine catch, which is largely meant for export, small-scale fishing feeds the domestic market, and a great amount of the catch is consumed locally, which means it requires less infrastructure and energy. Bag-net fishers also divide the year into periods where fishing activity intensifies or lessens, which gives a chance for fish to breed and grow.

When seen as fishery, the coast is not simply a static space where the many activities that constitute fishing happen. It is dynamic sea and landscape produced through these activities and its allied institutions and markets. Each kind of fishing produces the coast differently and entails a different horizon in which human-nonhuman relationships unfold. Just as small-scale fishing, trawler and purse-seine netters too produce fisheries, fish, and the fishing community. However, unlike small-scale fishing, which happens through deliberately cultivated interdependencies with fish where the coast is not treated as a predetermined terrain, intensive fishing practices create a coast that is a set field with a quantum of catch. The Department of Fisheries of Maharashtra makes this clear through not just its support for intensive fishing, but also through its ever-repeating refrain of "missed potential" (Bavinck and Johnson 2008): the idea that there is catch out there that is annually missed. This anthropocentric, growth-based model organizes the fisheries industry to keep increasing the annual catch by capturing greater resources and injecting infrastructure and revenue into intensive practices and technology. The consequences of large-scale fisheries and fisheries infrastructure are evident in their effects on coastal ecologies and communities, rising conflict between small- and large-scale fishers, and the sharp decline in fish stocks (Bavington 2011; Degnbol and McCay 2007).

The CRZ is critical to this growth-based model because it demarcates the terrain in which the fisheries industry operates and the area where resources are mobilized and managed. For the CRZ policy to operate, it has to first delineate what constitutes the coast—that is, the beginning and end of the area that is classified as the coastal zone. For this, the CRZ maps two lines: the seaward extent of what constitutes the coast, and its landward extent. Each of these lines is mapped in a different way and reconstitutes the fisher community differently. On the seaward side, where the extent of the coast is the extent of the territorial waters, the fishers are cast as sentinels who, after incidences of terror attacks from the seas (and incidences like MV *Pavit*, which we saw at the beginning of this chapter), are part of a coastal defense program to identify possible threats.[15] On the landward side, they are both stakeholders and knowledgeable wardens of the coast who are also dependent on it for their livelihood. The process by which the CRZ delineates the coast and designs it as a fishery can be thought of as a way in which it is simultaneously "localized, universalized, and naturalized to legitimate management" (Moore 2012), which, in turn, supplies the available terms such as *stakeholders*, *catch*, *vulnerable species*, and *bycatch* that refigure human-nonhuman relationships at the shore.

Because they are less intensive, small-scale fishing practices bring in a smaller total annual yield. In the fishing season between 2010 and 2011, nearly 500,000 tons of fish were caught in the state of Maharashtra. Of this, 143,000 tons of fish were caught along the shores of Mumbai, which accounts for just under 30 percent of the state's total production. The small cluster of fishing villages that Malvani belongs to, accounts for just under 3 percent of the entire state's yearly catch.[16] According to the Maharashtra State Department's fisheries estimates, each mechanized boat made an average of ninety-one efforts in that year—the term *effort* is defined by the number of times a boat casts and hauls its nets. This statistic can be read in different ways: it can bring home the precarious lives that fishermen lead and the tight margins in which they operate, since each mechanized vessel only ventures out once in three days, making it all the more important to maximize the potential of every catch. This is certainly the line that the Maharashtra Fisheries Department takes as it sees the continental shelf as an economic zone that has a specific quantum of potential to be tapped each year.[17] This understanding is a legacy of the Blue Revolution—the Indian state's project to inject technology and modernize fisheries that began in the 1950s onward (Bavinck and Johnson 2008). The Indian government began a systematic plan to support trawler fishing at the national level, whereas the support for mechanizing relatively smaller boats came at the state level (for instance, Maharashtra began providing support to mechanize boats from 1954 onward). Such initiatives framed the nearshore as the domain of the "local" where artisanal fishing happens and the offshore, deeper waters where the trawlers plied as the domain of the translocal, setting new spatial imaginaries of the coast (Subramanian 2009).

The ways in which official statistics keep count of catch versus "effort" and fleet size does not consider the complex sociocultural and economic arrangements that define small-scale fishing practices (Ota and Just 2008). As a result, they are often described as inefficient or in terms of lack. However, the other way to think about these statistics is that it reflects an inherent aspect of small-scale fishing. The coast and its fisheries that intensive fishing practices create are not the same as the coast of the small-scale fishing. Thus, this statistic—that small-scale fishers only make ninety-one efforts a year on average—should be taken as a historical, technological, social, creative process of bringing the coast, as sustainable fisheries, into being. Unlike the coast as zone of quantified potential, small-scale fishers create the coast through interdependencies that work in the long and short term. This should not be taken to mean that small-scale fishing lacks technological or economic sophistication or that numbers and quanta are irrelevant. Instead, it should be taken to mean that the ways in which small-scale fishing organizes the coast as a spatial entity and the practices it employs work through relationships and rhythms of nonhuman entities such as fish breeding cycles and tides, and it focuses on collective sustenance and resource management. The asymmetric distribution of opportunities and catch notwithstanding, small-scale fishing offers the opportunity for critical reflection—for finding ways of producing a more sustainable coast in the long term. However, these diverse practices and marine tenure arrangements are kept out of both coastal policies and coastal plans, which only approach them via the question of rights for traditional communities.[18]

Fixing the Tide

While the CRZ fixed the seaward side of the coast as a territorial border, it fixed the landward side in a different way. Drawing the landward extent of the coast depended on surveying the high tide line (HTL) identified via aerial and field surveys. The HTL has always been the most important determinant of the CRZ. In the 1991 CRZ, the parts of the coast that would fall under its jurisdiction were defined as "the coastal stretches of seas, bays, estuaries, creeks, rivers and backwaters which are influenced by tidal action (in the landward side) up to 500 meters from the high tide line (HTL) and the land between the low tide line (LTL) and the HTL."[19] Thus, the implementation of the CRZ depended on identifying and demarcating the HTL, as the extent of the coast and its subzones are only comprehensible in relation to this line. For example, the landward extent of the CRZ is simply offset from the HTL as a parallel line at a radial distance of five hundred meters. In a report on the process of rehousing those displaced by the 2004 Indian Ocean tsunami, Arthi Sridhar (2006) writes that this manner of identifying the CRZ did not take into account any land features such as mountains or cliffs, but simply denoted the HTL as a line that was traced on a cartographic map as from a bird's eye view, thus flattening and disregarding topographic features.

Subsequent notifications have kept the "highest spring hightide line (HTL) measured over ten years" as the main line that determines the landward extent of the coast. The most recent CRZ notification too follows this convention where the landward edge of the coastal zone is simply offset from the HTL. However, the nature of mapping the HTL has undergone considerable change with changes in survey technologies. For the most part, high and low tide lines are established by examining high-resolution satellite imagery along with bathymetric and tidal records. However, the crew assigned the task of drawing the CZMPs undertook ground surveys to determine the high tide line.

To determine this line, the surveyors had to look for particular signs. For example, the point up to which the tide ingresses into land was not determined by looking at the water itself, but by the presence of vegetation, dunes, berms, and other elements such as flotsam, and permanent construction along the shoreline. However, as I walked with the government surveyors appointed to map the CRZ, it became evident that determining this line was not a straightforward process because, often, nonhuman elements would act or appear in ways that confounded the mapping process. The plants, animals, and nonhuman entities that make up the coast-scape resisted attempts to be mapped into the CRZ's classificatory orders and ran against its neat categories.

In 2012, I followed around surveyors from Anna University who were appointed to draw the CZMP for Mumbai by the Municipal Corporation of Greater Mumbai. The CRZ policy assumes that the availability of high-resolution surveys makes the process of drawing the CZMPs a relatively straightforward process that, given the advancement in survey technology, will result in an accurate image of the coast. However, as the previous chapter shows, accuracy is not an inherent property of technologies or of high-resolution images but is built in through different practices in the construction of the CZMP. In the case of HTL, ensuring accuracy involved field surveys and ground truthing, which in turn involved interpreting signs on the ground to create a line on a map.

I was following Senthil, a surveyor who was a part of the team from Anna University. We were accompanied by Mr. Narayan, a municipal engineer, and Jayesh, a daily-wage worker hired by the MCGM. Our task was to walk along the coast of Aksa, Marve, and Gorai to determine the HTL. On the first day, we started walking from beaches south of Aksa. Since the HTL is determined by the presence of vegetation and infrastructure like seawalls, Senthil would direct Jayesh to stand at the edges of these elements and take a reading. Jayesh, who was holding the heavy GPS equipment, would stand there for the time required to get multiple locational readings from the satellite and the fixed station. Senthil would oblige and point to things and say, "See this, see? You can tell from this line of plants that the tide comes up to here." He was carrying a high-resolution satellite image on which he would point out structures and then point to the corresponding structure on the ground—a wall in this instance—and explain how that was a marker of the tide. As the water could not come beyond the wall, it was an automatic determinant of the HTL.

Not all elements along the coast were, however, as cooperative. Senthil would often come across things that did not appear clearly in the aerial image, or the image did not match up with what was on ground, and this necessitated ground truthing. Ground truthing is a process of direct observation—of going and looking at mapped or visualized elements on ground—as a means of confirming (or rejecting) inferred observations based on visual interpretation. For instance, the aerial image showed several blue squares on the beach. Since blue tarp is often used for covering hutments, Senthil had assumed that these were residential structures, but was unsure given their proximity to the water and wanted to confirm exactly what they were. When we got to the beach, there were no blue squares to be found. A few meters ahead, Senthil spotted a line of blue plastic sheets on which fish had been set out to dry, and from this, he figured that those blue squares too were just temporary plastic sheets and not of any importance in determining the HTL.

SENTHIL POINTED TO A GREEN PATCH ON TOP OF A SANDY SLOPE. THIS, HE SAID, WAS A GOOD INDICATOR OF THE HEIGHT TO WHICH THE TIDE ROSE. ONCE MAPPED, HE WOULD USE THIS LINE TO DETERMINE THE LANDWARD EDGE OF THE COAST.

THIS PLANT DETERMINES THE END OF THE COAST?

SINCE THE POLICY CAME INTO EFFECT, IT DOES

During the process of fixing the HTL, elements would disappear and appear or behave in unforeseen ways. Further north of the beach, Senthil and the crew were following a green patch that had grown on a sandy part of the beach. Senthil was explaining that vegetation is a good indicator of the HTL as a well-established patch of green shows the height to where the water comes and would not have survived if it were regularly inundated with saltwater. As we walked further however, a second patch of green had spurted up, much closer to the water. Senthil walked toward it and tried to see whether he could spot it on the reference image, which he could not. After walking up and down for a while, he finally asked the engineer if there were any structures here that had been recently moved. The engineer could not provide a conclusive answer. Senthil kept peering at the reference image until he arrived at an explanation that seemed to satisfy him: he observed that there were boats parked in the image and guessed that perhaps these boats protected that little patch of sand from saltwater, which had allowed the plants to grow. In any case, he said, he could not take this patch as "permanent vegetation" and it made more sense to stick to the well-established line of vegetation as a marker of the tide. Mapping the HTL involved many such acts of reading and determining the coast. The hardest part to draw turned out to be the creek, where the line had to be measured not in terms of any visual evidence, but in terms of salinity.

In the 2011 CRZ, the creek presents as an interesting conundrum: since it is the point where seawater enters the landmass, it makes it hard to determine where to draw the landward end of the coast. The CRZ attempts to get around this problem by measuring salinity and tidal action:

> CRZ shall apply to the land area between HTL to 100 mt or width of the creek whichever is less on the landward side along the tidal influenced water bodies that are connected to the sea and the distance up to which development along such tidal influenced water bodies is to be regulated shall be governed by the distance up to which the tidal effects are experienced which shall be determined based on salinity concentration of 5 parts per thousand (ppt) measured during the driest period of the year and distance up to which tidal effects are experienced shall be clearly identified and demarcated accordingly in the Coastal Zone Management Plans.

Thus, the team at Anna University were to collect samples to measure salinity. However, when the field survey team did so, they found that their salinity meters were unable to give a proper reading because of the level of pollutants in the water. During a meeting where the team was giving the MCGM a quick progress report, the team said that these samples would have to be sent to be analyzed by an external lab to determine the extent of the coast along the creek.

To measure salinity, surveyors used a hand-held device called a refractometer. Refractometers use light to measure the concentration of substances dissolved in an aqueous solution. As light passes through the sample, it bends, and the meter measures the angle to which it bends to determine the concentration. Refractometers are used because they are cheap and portable and offer quick results, which makes them ideal for field surveys. However, they also have a higher degree of error and are not that sensitive (Patil and Shaligram 2013). The surveyors had to send the samples for chemical analysis because the substances in the coastal waters—pollutants—threw off the readings. Seeing, recording, keeping logs, taking GPS readings, refractometers, and chemical analyses comprise a range of "sensing practices" (Howe 2019) that allow the surveyors a way to understand specific environmental conditions, formations, or transformations—in this case, salinity. Paying attention to these sensing practices at the shore is also a way to understand the inseparability between the material world and the "formalisms we use to describe it" (Helmreich 2014, 266). These practices bend the presence or actions of nonhumans in the service of the CRZ's central task, which is to impose lines on the landscape to order and control it.

During my tenure with Dev and Gauri Koli's family, I was explained the process of recognizing and assimilating signs, pieces of information, and the successive decisions that guided fishing through close analogies with the creek. The creek, an inlet of water where the sea slips into the landmass, and where fresh water from lakes and runoffs fall into the sea, is an extremely important element in fishing and coastal life. The beginning and end of the fishing season is marked by changes in the creek's behavior, as it "churns" with the monsoon and is "quieted" after the rains. This behavior is also a sign to determine the depth at which to fish, as the creek and the seawater are thought of as a *mixture* that is either stirred or still. To the fishers, creeks, estuaries, and bays offer a measure of protection from the direct action of the sea, places to park their boats, dry their fish, and in times when the sea is rough, a place to fish. The creek, by virtue of its form and movement, sustains forms of life both in it and along its borders. Given that fishers have actively shaped the creek through the cultivation of mangroves, fishing infrastructure, and other construction projects, it can be thought of as a milieu in which the uncertainties, rhythms, risks, and possibilities of fishing circulate. The CRZ completely overlooks this point as it posits an unchanging relation between fishers and the coastal landscape. It assumes that the creek has an integral, unchanging form, and that its uses can therefore also be circumscribed by unchanging tradition. Instead of easily bending to the surveyor's eye, sites such as creeks resist and confound categorization. These sites can be points where the two milieus—of fishing and the CRZ—meet and produce a churning or agitation that amplifies anxieties around both long- and short-term concerns.

Possibilities that Rise Out of Muddied Waters

In recent years, several national and regional dailies published articles that warned the city's residents about the dangerous levels of heavy metal in fish caught off Mumbai's coast. These pollutants have a visible effect on marine creatures, with increasing algal, plankton, and jellyfish blooms that not only suffocate the fish, but also destroy the catch by rendering it inedible.[20] On the beach, it was not unusual to find the catch covered in blobs of purple and pink—a sign of a jellyfish bloom. Not only do the fishers have to spend a lot of time and effort in cleaning the catch to sell it at the market, but this laborious process also leaves them vulnerable to the effects of these pollutants. For instance, the process of manually separating and cleaning jellyfish (and the whole catch in general), which is a painful task, is done without any protective gear. Since fishers are also consumers of fish—the leftover catch is almost always taken home for domestic consumption—they are also most vulnerable to the effects of heavy metal poisoning. Artisanal fishing communities are also facing the impact of overfishing, which is a problem created partly by intensive fishing technologies such as trawling and purse-seine netting. As Alicia Fentiman's (1996) study of the impact of drilling for oil on fishing communities in the Niger Delta shows, pollutants and climate change stressors are not just experienced on the body, on ecology, or on livelihoods, they also leave deep marks on other aspects of political and cultural life.

To meet the challenges posed by pollution and climate change, not only do communities have to find adaptive strategies (such as recognizing signs of impending weather and changes in seasonal cycles among other things), but also invest in, and use, new technologies and tactics (Pinho et al. 2012). While the CRZ does attempt to address the long-term concerns of pollution and climate change, especially through massive infrastructural projects, the problems and anxieties it creates in the short-term prevent fisher communities from fully engaging with these ecological visions (even if to dispute them). It also delimits the forms of ecoauthority (Howe 2014) they can deploy in their political struggles to retain access to land, commons, and fishing infrastructure while ensuring they do not miss out on the developmental opportunities that the policy offers.

I had arrived in the field in May 2011 when the fishing season had already begun winding down. I began by familiarizing myself with the 2011 CRZ and its land politics. The problem was that the 2011 CRZ had opened the city's coast for development. However, the fisher's right to develop their settlements depended on their ability to prove that they were, in fact, traditional communities. The result was that the fisher community now faced the increasing possibility of displacement—gaining control of the development potential of their lands remains one of the central goals of the community's political activism since 2011. While environmental concerns are also important to the fisher community's cause, there is very little written or spoken about the particularities and the mechanics of nearshore fishing, the web of relationships and forms of interspecies care that sustain it, and the ways in which it creates and sustains the coast.

ANOMALOUS LANDSCAPES

Infographics, Extreme Weather, and Urban
Infrastructure

Blip

The environmental engineer, who was sitting across the table from me, said:[1]

> How can I think of this as a river? I mean . . . it is called a river . . . but it is not so simple because a "river" means something—and it is a very complex entity. It is brackish on one side—you know, saltwater comes in; there are mangroves and mudflats, which flood. For some time during the year, it is dry. In some places, it looks like a gutter. By that I don't mean that it is small; I really mean it looks like a gutter or a drain, with concrete sides and filled with sewage and garbage. So, I faced an interesting question—how do I think of this as a river? I know this sounds confusing, but what I'm trying to say is that this river [the geographic entity] and that river [the river in the report] are not the same rivers.

Dr. Kumar was a part of the team tasked with making recommendations to improve the conditions of the Mithi, Mumbai's longest drainage channel, in the aftermath of the 2005 Maharashtra floods. On July 26, almost a meter of rain fell in Mumbai over a period of twenty-four hours and resulted in the death of more than a thousand people. Though, eventually, 2005 proved to be a year when the monsoon underperformed, the intense rainfall on July 26, 2005, stands apart as an anomalous, unprecedented event. This is evident in the statistical images that circulated after the event—graphs that contrast the volume and intensity of the precipitation on that day with average highs during preceding seasons. These images reshaped the city's landscape and its governance. The impact of this statistical blip can be seen in the changes to the city's drainage systems, of which the Mithi is an important part.

Apart from the spectacularly high precipitation, the flood was also seen as the inevitable outcome of the city's characteristically dysfunctional infrastructure. To prevent future floods, the Municipal Corporation of Greater Mumbai (MCGM) set out to restore the city's "natural" drainage channels and, consequently, launched a massive river restoration project. The MCGM's postflood emphasis on restoring the city's rivers is curious given that the city's "primary ecology," as Anuradha Mathur and Dilip Da Cunha (2009, 20) show, is that of an "estuary in the monsoon" where "unlike deltas where rivers reach into the sea, estuaries allow the sea in." This is not to imply that estuarine systems and riverine systems are wholly separate, rather, to say that the MCGM's project imagined a particular kind of engineered landscape for which it operationalized the term *river*. Second, as in many other Indian cities, the drainage system is a "socionatural assemblage" that defies categorization (Ranganathan 2015). Mumbai's drainage system emerged over different eras of the city's construction, and is a complicated mix of underground drains, roadside gutters, ponds, and concretized channels that connect with creeks and mangroves that meet the sea. Thus, what did this rediscovery of the city's rivers signify and where do we find the genealogy of this invented category?

The answer to these questions lies in the emerging concepts of how everyday urban life, landscape, and infrastructural and governmental systems must change in a time of extreme weather. These material changes emerge from visualization of weather anomalies as blips in relation to a stable weather of the past.[2] In Mumbai, a wide fascination with tracking rainfall data through social media emerged after the 2005 flood along with a meteorologically minded online public organized around reading, creating, and circulating weather data. These graphs and statistical charts place day-to-day experiences in relation to longer patterns. The publics they organize show how residents do not just think of the weather as separate from the city, but in terms of a city-weather assemblage.

The images are a means for addressing "meteorological anxieties" (Amrith 2018) across scales: for making everyday decisions—whether to stay home, whether to leave work early, and what routes to avoid during heavy rain. They are also used to justify large-scale urban interventions to fix the hypervisible infrastructural dysfunction and meet the challenges posed by climate change. Projects like the postflood river restoration are examples of how landscape and infrastructure are reimagined in coastal cities across the world to meet the anticipated risks of the climate crisis. Weather visualizations are important media through which this rescaling happens.

In post-2005 Mumbai, the rising incidence of heavy rainfall introduced new expectations about the carrying capacities of urban infrastructure, that is, the volume of rainfall that the drainage system must be capable of handling without spilling over. To meet these volumes, it was necessary to secure the identity of Mumbai's hydrological landscape as riverine. This meant widening and standardizing the drainage system; training the expansive, meandering, and fickle estuarine systems; and turning wide marshy areas and creeks with blurred edges into deep riverine channels with defined, concrete edges.

Thinking of Mumbai's drainage as a system of rivers meant turning away from the expansive and often paradoxical surface area logic of estuaries and heterogeneous systems to the volumetric logic centered on what could be engineered as standardized concretized channels and called rivers. The focus of this project was the Mithi, as many attributed the damage caused by the flood to its deteriorated condition, choked as it is with urban detritus. While the Mithi is often described as Mumbai's longest river, it is also described as a drainage channel, a gutter, or a creek. Dr. Kumar's words at the beginning of this essay illustrate this clash between the "Mithi River" that the study ushered into being and the "river" that he was referring to: a mixed bag of systems, waters, and infrastructure that crisscrossed the coast.

Mithi's new identity shows how the circulation of meteorological information as statistical figures and infographics shapes the public understanding of extreme weather events and disaster management. Data visualizations and meteorological infographics—visualizations that communicate a narrative using different kinds of images—are not just about new perceptions of the weather but also set in place an aesthetic of objectivity that legitimizes large-scale urban projects. Though very different in nature from survey and plans, they are representational projects that reorganize life (Halpern 2015) and bring into being a different kind of an urban, coastal landscape. They are entangled with climate change politics, postdisaster governance, the history of drainage infrastructure and rainfall prediction, and urban land politics in Mumbai. The graphs of changing rainfall patterns are integral to sustaining the idea of a riverine Mumbai and act as an important conduit for mixing postdisaster rhetoric of environmental responsibility with future infrastructural endeavors. However, these images also have other consequences: the process of concretizing the riverbanks also became a means of displacing informal settlements, reclaiming marshes, capturing land for real estate development, and further entrenching the politics of belonging that divides Mumbai.

THE SCALE OF THE 2005 FLOOD IS PALPABLE IN THE ACCOUNTS THAT ENUMERATE NOT JUST THE DAMAGE, BUT ALSO THE RECOVERY EFFORT. THESE STATISTICS AND NUMBERS WERE IMPORTANT WAYS IN WHICH PEOPLE MADE SENSE OF THE EVENT, DESPITE THE FACT THAT THEY WERE COLLATED AT DIFFERENT TIMES, SOME LONG AFTER THE FLOOD HAD PASSED, AND DESPITE THEIR VARIATION ACROSS SOURCES.

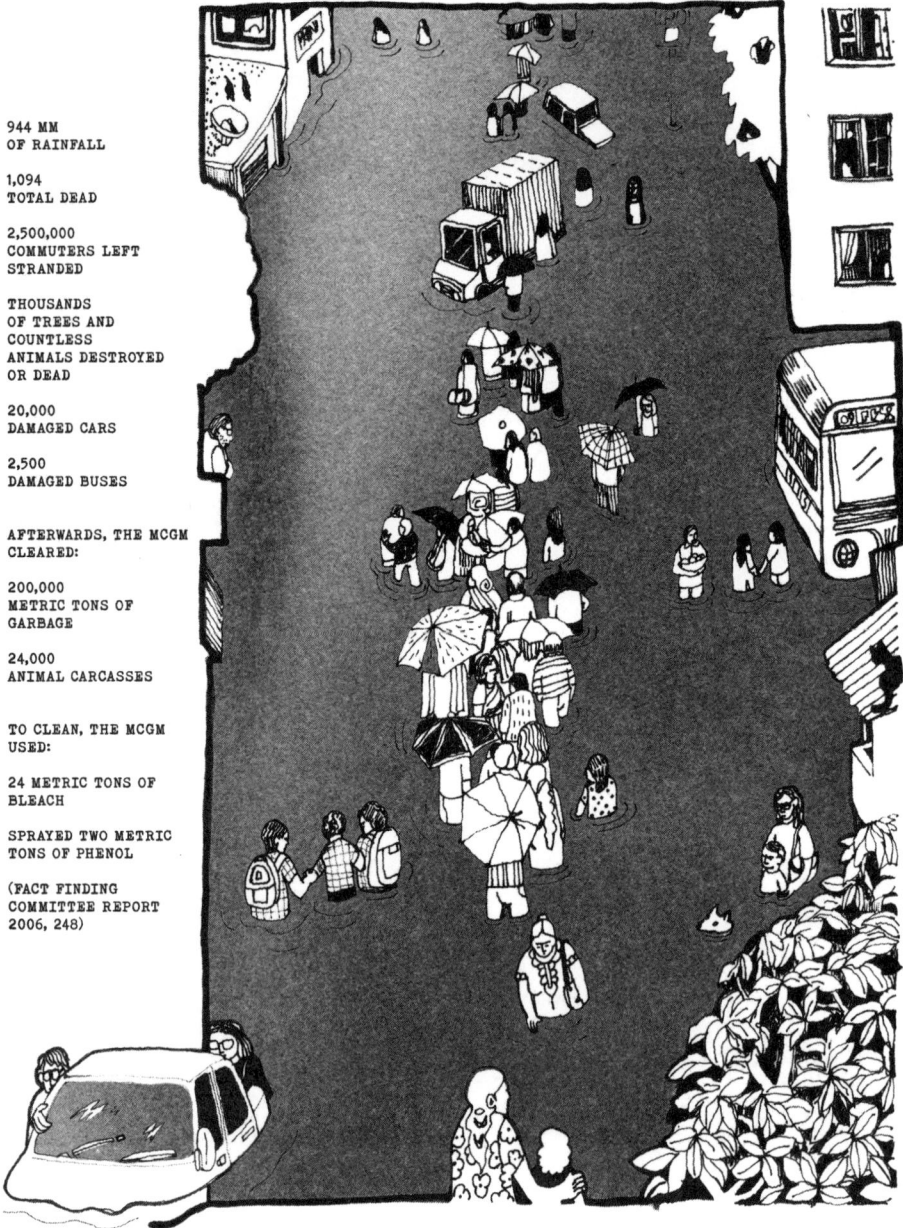

944 MM
OF RAINFALL

1,094
TOTAL DEAD

2,500,000
COMMUTERS LEFT
STRANDED

THOUSANDS
OF TREES AND
COUNTLESS
ANIMALS DESTROYED
OR DEAD

20,000
DAMAGED CARS

2,500
DAMAGED BUSES

AFTERWARDS, THE MCGM
CLEARED:

200,000
METRIC TONS OF
GARBAGE

24,000
ANIMAL CARCASSES

TO CLEAN, THE MCGM
USED:

24 METRIC TONS OF
BLEACH

SPRAYED TWO METRIC
TONS OF PHENOL

(FACT FINDING
COMMITTEE REPORT
2006, 248)

Cloudburst

On July 26, 2005, almost a meter of rain fell in Mumbai over a period of twenty-four hours and the flood claimed more than a thousand lives. Due to the tremendous scale of the rain and the destruction it caused, the 2005 flood is deeply embedded in urban memory. It resurfaces annually with each monsoon and has become an archetype of the extreme weather events associated with anthropogenic climate change. This is also evident in the postflood scholarship on urban planning and climate change (Anand 2017; Mathur and Da Cunha 2009; Rademacher 2011). Amitav Ghosh (2016), who, like Mathur and Da Cunha, relates the flood to the destruction of the estuarine systems on which Mumbai was built, uses the event as an inflection point in meteorological thinking in the city. He writes that as meteorological events can no longer be predicted and prepared for as before, and because of their intensity and the quickness with which they arrive, they create new forms of anxiety about the weather.

Apart from the scale of destruction, information on the flood and its circulatory networks created new anticipatory anxieties, which in turn changed the sense of risk that came to bear on postflood urban initiatives. The anxieties were directed both toward the inability of the city's storm water drainage to handle large volumes of precipitation and toward a possible return of the enormous rain. While there was little information about the flood as it happened, a few days afterward, a narrative emerged that was vital to constructing a collective sense of the event that had just passed. Newspapers began describing the meteorological event as a "cloudburst," a phenomenon where a large cumulonimbus cloud gathered over the city and dropped all its water at once. This giant volume of water fell on a city with a neglected, garbage-choked drainage system constructed on an ad hoc basis (Times News Network 2005). The resulting damage was severe because not only had the weather monitoring stations failed to predict the event, but also because there was no information broadcast before or during the event (Municipal Corporation of Greater Mumbai 2006). This narrative helped explain the flood as an amalgamation of a meteorological phenomenon, failing infrastructure, and lack of communication. However, the narrative needs to be unpacked in relation to the new life of meteorological communication and visual representations of the weather because they not only tell the story of the event, but also impact the infrastructural responses that follow.

Terms such as *cloudbursts* create a social imaginary of meteorological events, and in contemporary times, shape how we understand and anticipate the increasingly volatile and intense weather of the Anthropocene. However, as histories of meteorological communication show, a cloudburst is at best a nebulous term; it is as much a media phenomenon (Harris and Lanfranco 2017) as a meteorological one. Though it has been in use since the nineteenth century, it was only in the 1990s that the term became associated with a threshold value that separated it from other kinds of rain, which gave it some semblance of having a concrete definition. Through its usage in research that studies flash floods and intense downpours in the Himalayas, it came to be defined as rainfall that exceeded 100 mm/hour, a measure that was adopted by the Indian Meteorological Department (IMD). Moreover, it is a term that is applied to a phenomenon in hindsight, after the event has passed, rather than a phenomenon that is predicted, such as a cyclone (Harris and Lanfranco 2017). Despite its murky past, the term *cloudburst* changed significantly after 2005. The MCGM deployed the term in its explanation of why critical urban systems had failed: since cloudbursts were unprecedented events that rarely happened, urban systems and infrastructure were not designed, nor could they be designed, to cope with the volume and intensity of precipitation that cloudbursts generate. Ironically, the fact that the IMD had failed to predict the cloudburst on July 26 later became a bone of contention.

In addition to the cloudburst, an important factor for the public fascination and the collective trauma from the flood was the lack of information that was available before or during the event. The IMD had predicted "heavy to very heavy rainfall," which, according to its official categories, corresponds to 115.6 to 204.4 mm of rain (Municipal Corporation of Greater Mumbai 2006). Soon after the rain began, phone lines and cell phone communication stopped working and there was no information forthcoming from any civic institution (Municipal Corporation of Greater Mumbai 2006). The Fact Finding Committee on Mumbai Floods, which was created soon after the event, describes how this data gap was one of the key reasons why so many systems failed on that day, leading to a high number of deaths. It set off a cascade of events: the IMD failed to predict the level of precipitation or alert the city's disaster management unit as the cloud formed, vehicular traffic continued to ply on waterlogged streets creating enormous jams, there was no way to communicate and coordinate emergency response, and as the day wore on and the rain did not stop, people began to panic (Fact Finding Committee Report 2006).

In the immediate aftermath of the flood, reporters and researchers approached the IMD's regional office in Mumbai to gather rainfall data, which were scant since the flood had submerged the rain gauges.[3] The accounts from right after the flood were mostly photo driven, such as pictures of commuters wading through flooded streets, people struggling in neck-deep water, and rescuing trapped residents with makeshift rafts and boats. These were followed by accounts of damage in low-lying areas and stories of urban development gone wrong. Diagrams describing the effects of the event at the scale of the city took longer to come. One of these was an image with an outline map of the city that showed the flooded areas and a schematic diagram of how a low-pressure vortex led to a cloud-burst. Data collected over the next few months came in the form of hydrographs (rainfall readings), tide tables, water levels in the city's various catchments, and much later, maps of flooded areas. There was no direct route through which these data circulated; parts of information in research reports would then circu-late in newspapers, which meant that public information about the flood came from varied sources at different times. Representations of data that circulate after disasters, Max Liboiron (2015) writes, are crucial to apprehending the experience of extreme weather events and this was also the case in the 2005 flood. Despite the dispersed nature of their circulatory networks, diagrams, graphs, and maps build "data imaginaries" that helped make sense of the collective experience of the event (Liboiron 2015).

All these different images constructed the why, what, and how of the flood, giving it narrative coherence where an unprecedented phenomenon (cloudburst) interacted, by chance, with another phenomenon (the tide). It brought attention to the material conditions of the deluge, the uncontrolled mixing together of salty tidal water with the rainwater in an unregulated drainage network made worse by the glut of trash. However, just as this picture emerged from data, it also brought attention to missing data: the lost automated readings, the failure to pre-dict the right intensity of rainfall, and the failure to interpret and communicate weather data. Much as with the problem of the garbage in the drains, the lack of channels for data to flow between institutions and the inability to respond to data in a timely manner caused as much, if not more, damage as the water itself. July 26 was later described as a day that the city came to a standstill and the gaps in the data amplified the sudden break caused by the event.

These informational gaps generated a public and institutional interest in rain-fall data, which has risen in recent times with the increasing frequency of floods. The shifts in monsoon reportage in newspapers shows how it is not just data that fuels this meteorological interest, but also the ways in which those numbers are imaged and juxtaposed along with other images to create infographics—compound visual representations designed for communicating technical infor-mation as a narrative to a wide audience.[4] In the context of the monsoon, these infographics often present different packets of technical information that com-bine statistical visualizations and maps to create a narrative about shifting rain patterns.[5] The main components of the monsoon infographics are statistical figures and several different statistical figures are strung together to create par-ticular narratives about flood or water scarcity and about the annual experience of the monsoon. Narrativized infographics allow for a colloquial understanding of technical data, and this is not just limited to floods in Mumbai, but to the relationship between images and the public understanding of climate change more generally (Schneider and Nocke 2014).

The 2005 flood had shown the need for weather communication and for local, real-time rainfall information, which over the course of subsequent monsoons became not just easier to access, but also the focus of the daily news about the monsoon. While predicting, following, and worrying about the monsoon has a long history, both the increasingly erratic rainfall and new forms of data, particularly statistical data, have changed the way people worry; it has changed how they have arranged their lives in relation to short and long-term predictions of rain and their relationship with the city and its systems.

JACQUELINE URLA (1993) WRITES THAT STATISTICAL FIGURES PRODUCE "TRUTHS" IN COMPELLING WAYS; THEY "OPERATE SIMULTANEOUSLY AS TECHNOLOGIES OF SCIENTIFIC KNOWLEDGE, OF GOVERNMENT ADMINISTRATION, AND OF SYMBOLIC REPRESENTATION." INDEED, THE LACK OF STATISTICAL DATA AFTER THE FLOOD AMPLIFIED THE PUBLIC MISTRUST IN THE MCGM AND THIS ALSO EXPLAINS WHY THERE WAS A GREAT DEAL OF INTEREST IN FINDING DATA TO "TRUTHFULLY" RECONSTRUCT THE EVENTS OF THAT DAY.

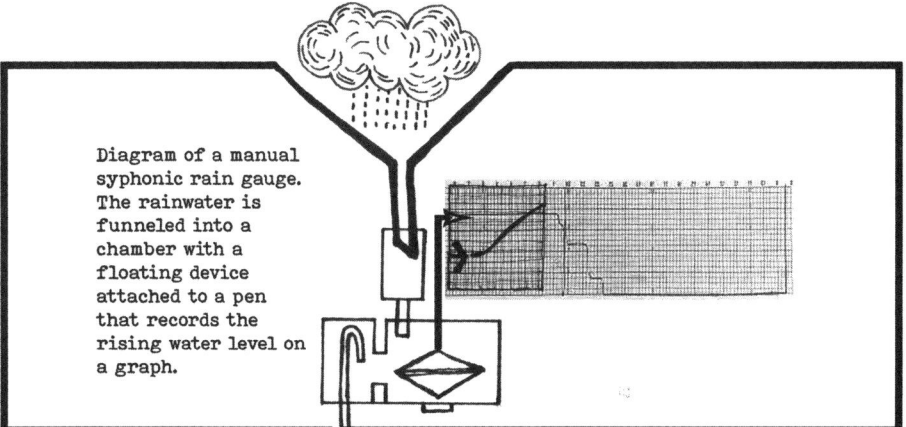

Diagram of a manual syphonic rain gauge. The rainwater is funneled into a chamber with a floating device attached to a pen that records the rising water level on a graph.

THE FACT FINDING COMMITTEE REPORT (2006) IS THE MOST COMPREHENSIVE DATA SOURCE ON THE FLOOD AND IT ACKNOWLEDGES THE LACK OF INFORMATION ABOUT RAINFALL, WEATHER DATA, AND COMPREHENSIVE DATA ON THE DRAINAGE SYSTEM. TO GIVE A SENSE OF THE RAINFALL DISTRIBUTION, THE REPORT UTILIZED DATA FROM THE SANTACRUZ WEATHER STATION, WHICH WAS ONE OF TWO WEATHER-MONITORING STATIONS OPERATIONAL IN 2005.

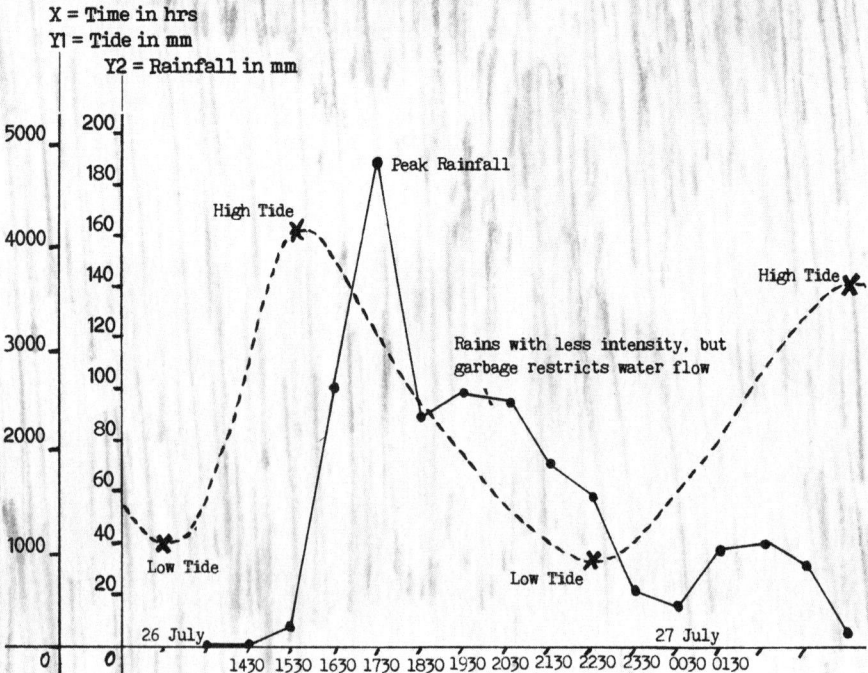

IT NOTES THAT THE AUTOMATED RAIN GAUGE, THE MOST SOPHISTICATED
TOOL FOR RECORDING RAINFALL DATA AT 15-MINUTE INTERVALS, SUBMERGED
PRETTY SOON AFTER IT STARTED RAINING AND COULD NOT BE USED AS A
DATA SOURCE. IN THE ABSENCE OF THIS INFORMATION, THE EVENT COULD
ONLY BE CONSTRUCTED USING HOURLY DATA FROM THE MANUAL RAIN
GAUGES. WHEN PUT TOGETHER WITH TIDAL INFORMATION, IT REVEALS HOW
THE FLOOD HAPPENED IN TWO WAVES WHERE THE HIGH TIDE HELD BACK THE
FLOODWATERS IN DRAINS CHOKED BY GARBAGE. THE RAIN GAUGE READINGS
AND THE TIDE TABLES WERE CRITICAL IN CONSTRUCTING A TIMELINE AND
IN ESTABLISHING THE SPECTACULAR SCALE OF THE EVENT.

X = Time in hrs
Y1 = Tide in mm
Y2 = Rainfall in mm

Peak Rainfall

High Tide

High Tide

Rains with less intensity, but
garbage restricts water flow

Low Tide

Low Tide

26 July

27 July

1430 1530 1630 1730 1830 1930 2030 2130 2230 2330 0030 0130

THESE STATISTICAL FIGURES WERE COMBINED WITH OTHER DIAGRAMS AND MAPS: *OUTLOOK* MAGAZINE PUBLISHED A DIAGRAM SHOWING HOW THE RAIN WAS A RESULT OF A LOW-PRESSURE SYSTEM THAT CREATED A VORTEX OVER THE CITY.

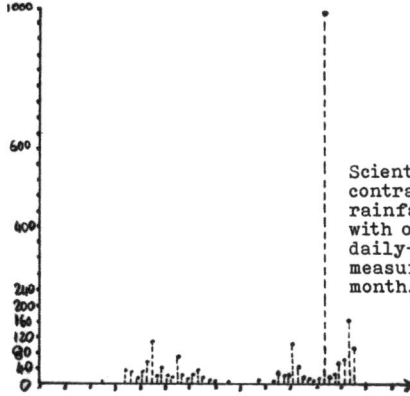

Scientific graphs contrasted the rainfall on July 26 with other daily-rainfall measurements over a month.

MARSH

FOREST

BUILT AREA

Maps of unchecked urban development circulated, linking the rain to the sociomaterial and infrastructural city.

1967

2005

The airport, which was badly affected, was an oft-cited example of bad planning practices as the Mithi had been diverted to make way for the runway.

THE FIGURES CAME FROM DIFFERENT SOURCES AT DIFFERENT TIMES TO CREATE A PERSISTENT NARRATIVE OF INFRASTRUCTURAL AND ADMINISTRATIVE FAILURE THAT COULD PROVE DANGEROUS GIVEN THE CHANGING WEATHER PATTERNS.

City-Weather Assemblage

Most of India's rainfall occurs between June and September each year. The IMD defines the monsoon as the "seasonal reversal of wind direction along the shores of the Indian Ocean, especially in the Arabian Sea, that blow from the southwest during one half of the year and from the northeast during the other."[6] Uncertainty, as Sunil Amrith (2018) writes, is intrinsic to the monsoon, even though it is characterized as a "season" that returns annually. Because the monsoon is a complex phenomenon affected by changes in ocean temperature and transoceanic currents, it has long eluded meteorologists seeking to forecast it accurately. Far from evenly distributed, the rain comes down in short and intense bursts, which further complicates predictions and resists any broad-stroke description. This characteristic uncertainty produces what Amrith calls "meteorological anxiety" for both the forecasters and for those who depend on regular rain for their livelihood.

While the postflood fears about intense rain and failing urban systems flow from this long history, anthropogenic climate change and its effect on the weather have amplified the risks and uncertainties associated with climate-induced disasters. The experience of monsoon too has shifted: studies show that rainfall in central India moves between extremes, causing repeated waves of flooding and drought (Roxy et al. 2017). In the postflood climate, where erratic weather is increasingly common, the nature of meteorological anxiety has changed as well. Even as patterns shift in dramatic ways that defy predictive capacities, people have come to rely on easy-to-access weather information, especially easy-to-read visual information, to make sense of this shifting experience. As Birgit Schneider and Thomas Nocke (2014, 9) write of the visual language of graphs and in the IPCC Reports, "Images have started to shape the imagination of a world under the conditions of climate change." These images are not so much about creating certainty or surety in a time of climate change, but rather about explaining the inherent uncertainty and limits of climate science and prediction—about giving shape to the meteorological unknowability in the Anthropocene.

The problem of meteorological anxiety after the 2005 flood has also to be understood in relation to the intervals at which the monsoon predictions are made.[7] Every year, the IMD produces two forecasts for the monsoon season, one in April and another in June, closer to the arrival of the southwest monsoon. Both these reports give a broad prediction of the gross rainfall for the season. Apart from these, the weekly reports, and the various other regional reports that supplement them, the IMD's Mumbai office puts out a rainfall reading every day at 8:30 a.m. This daily broadcast for the day was, until quite recently, the main source of information for Mumbai's residents against which the made everyday decisions. The 2005 flood changed that because the cloud-burst demonstrated how rain is a highly localized phenomenon and the asymmetric ways in which rainfall affects different areas given the heterogeneous nature of the city's drainage system. That is to say, the weather is not a thing by itself that happens to the city.[8] Rather, to understand the emerging practices of communicating and following the weather, it is helpful to think in terms of a city-weather assemblage—the weather as it interacts in different ways through the complex material objects, landscapes, and social relationships that make up the city.

In Mumbai, the 2005 flood stands as a marker that divides the meteorological uncertainties of before from the present-day anxieties, which, as amateur weather followers and municipal engineers would tell me, are about shorter-term, localized precipitation and its effects on the city's systems, which vary vastly across its different neighborhoods. The flood largely spared the Island City but had a devastating effect in the suburbs. It was particularly bad in some areas such as the airport and the suburbs of Kurla and Sion, which, as the Fact Finding Committee (2006) pointed out, are low-lying areas that were filled up without any oversight from the state planning authorities. The second problem was that of the intervals at which rainfall is reported: even if the city's two official weather stations—the Colaba and the Santacruz stations—broadcast the level of precipitation every morning, this is often of no help to people struggling to commute during heavy rain in the afternoon hours as the rain happens. Rain can fall in intense short bursts that are contained within a small area, and as traffic and rainfall have dramatically altered in recent times, this has created a need for weather information at shorter intervals that is relevant to commuters as they move across different neighborhoods. Since IMD did not broadcast localized data, groups of amateur meteorologists stepped in to fill this gap, forming loose collectives that gathered rainfall data to combine with other kinds of urban data at the level of the neighborhood.

As Internet platforms such as Twitter and Facebook became popular in the mid-2000s, small groups of amateur meteorologists began using them to circulate local weather data, adding to the regular forecasts put out by the official channels. In interviews and on their blog posts, amateur weather watchers in Mumbai cite the July 26 flood as the reason why they felt the need to transmit meteorological data via social media. While the IMD and the MCGM both have active Twitter accounts, the amateur weather watchers fill an important gap by providing localized real-time data. These official and amateur channels work in tandem along with several institutions and draw on official information. Amateur weather followers spend a considerable amount of time rendering meteorological data into statistical figures, graphs, and maps to make them more accessible to a wide audience. As these different bits of data circulate, they are combined with other information such as traffic updates and train timetables. This shows how contemporary meteorological anxiety in Mumbai is not just directed toward the weather alone. The combination of graphs, statistics, and information about the effects of the rain on the city attract and organize a public that is keenly interested in the city-weather assemblage.

One of the most important unofficial channels is a groups called Vagaries of the Weather, an Internet-based collective that reaches its audience via Twitter, compiling and transmitting data on the western region of the country, with a focus on Mumbai. Vagaries began as a weather blog authored by Rajesh Kapadia, a resident of Mumbai, shortly after the 2005 flood. Other authors joined in as the blog grew in popularity and apart from regular updates on the weather, many of their posts focus on temperatures, rainfall, and other weather-related events that vary from the norm. As Nitin, one of the contributors explained to me, this interest in variations has two dimensions: it shows how weather, especially rain, can manifest differently across the city, creating different microclimate pockets. Second, he said, many of his fellow weather followers wanted to focus on these variations to bring public attention to the problems caused by climate change.

The amateur networks work in a highly coordinated manner. Many of them work on specific regions and even within those, oversee specific tasks. They first go through official websites such as they IMD's webpage and that of the State Department of Agriculture to gather weather data derived from their sophisticated systems. Data on these sites, Nitin explains, are not published in a format that is easy to read. They are sometimes buried as links to PDFs, or they are put up in dense excel sheets, or they are presented in a manner that makes them inaccessible. For example, after the 2005 flood, the MCGM installed more than sixty automated weather stations (AWS) in the city, and the data they record are automatically sent to the MCGM and is publicly available on the Internet. However, these data are only accessible either through a cumbersome web portal or through a separate application. To reach a wider audience, people like Nitin regularly comb different official websites, download, and sift through the data, and then work on them; they create easy-to-read data tables and combine textual explanations, and most important, they often publish timely data such as rainfall records on a map. Nitin sees the regular task of visualizing data as one of the most important aspects of the social service they provide to the city. Maps and graphs make data not only easy to understand, but as he says, they make it possible to quickly read and process this information on the go.

THE AMATEUR METEOROLOGISTS AND THEIR ONLINE COMMUNITIES RESPOND TO THE VAGARIES OF THE WEATHER AS IT MOVES THROUGH THE URBAN ASSEMBLAGE THAT IS MUMBAI. THEY ARE SHAPED BY THE METEOROLOGICAL ANXIETIES OF CLIMATE CHANGE AND ITS INEXTRICABILITY FROM THE STUFF AND THINGS OF MUMBAI: A CITY OF 22 MILLION WITH OVERBURDENED INFRASTRUCTURE.

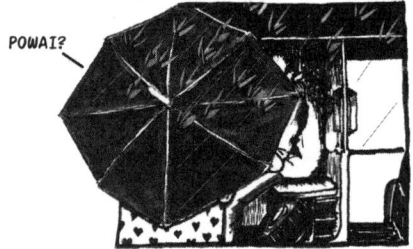

ARE TRAINS STILL RUNNING #MONSO

ANOTHER 200 MM RAIN RESPITE AI FLOODS

SAHAR COMPLEX WATERLOGGED #MUMBAI RAINS

FLOODING IN KHARGHAR #STAYATHOME

AUTO!

POWAI?

SEEPZ SE LENA SAKINAKA BHAR GAYA HAI. [GO VIA SEEPZ, SAKINAKA IS FLOODED]

NO KISSING

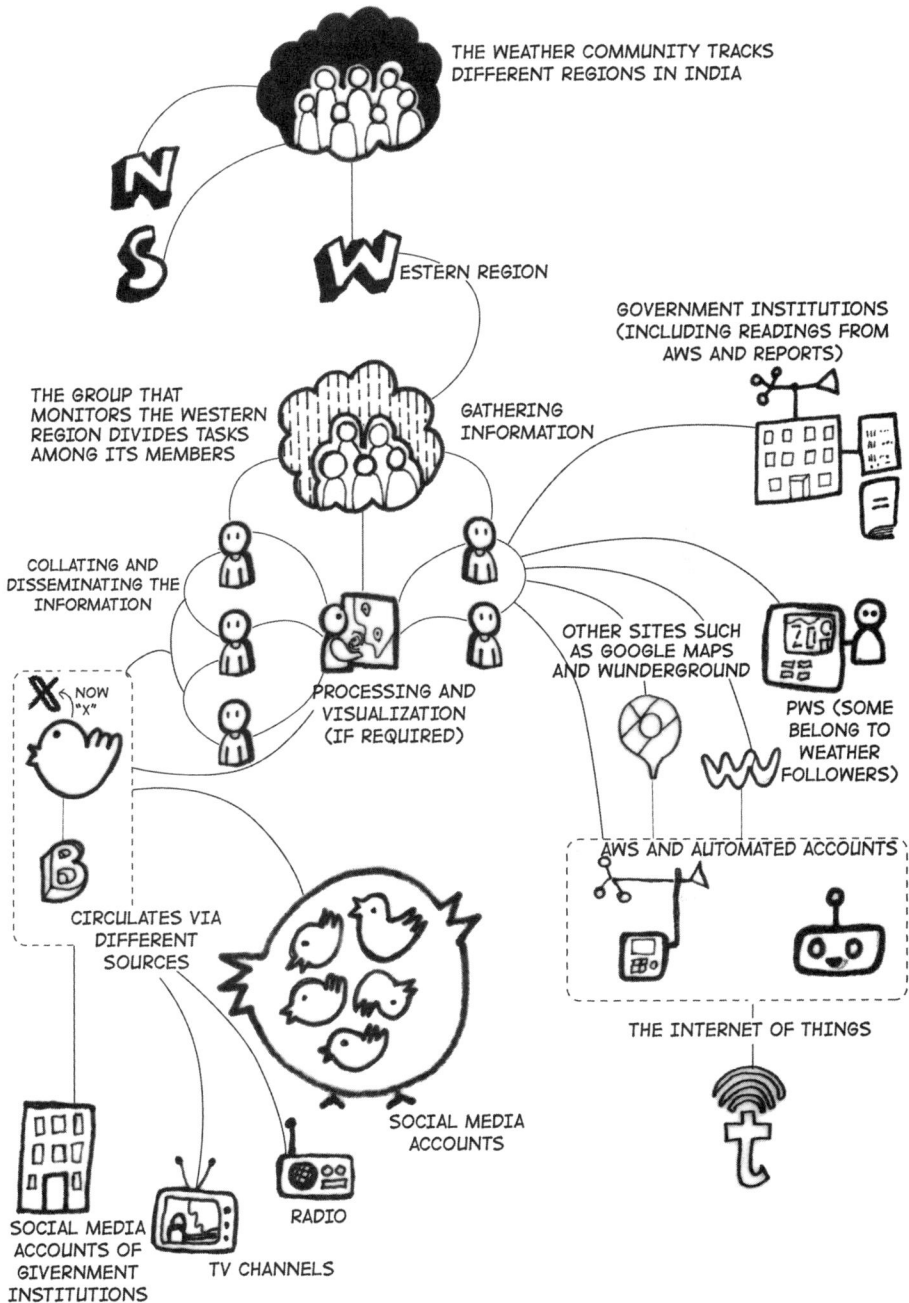

THE WEATHER COMMUNITY TRACKS DIFFERENT REGIONS IN INDIA

N

S

WESTERN REGION

GOVERNMENT INSTITUTIONS (INCLUDING READINGS FROM AWS AND REPORTS)

THE GROUP THAT MONITORS THE WESTERN REGION DIVIDES TASKS AMONG ITS MEMBERS

GATHERING INFORMATION

COLLATING AND DISSEMINATING THE INFORMATION

PROCESSING AND VISUALIZATION (IF REQUIRED)

OTHER SITES SUCH AS GOOGLE MAPS AND WUNDERGROUND

PWS (SOME BELONG TO WEATHER FOLLOWERS)

X NOW "X"

B

CIRCULATES VIA DIFFERENT SOURCES

AWS AND AUTOMATED ACCOUNTS

THE INTERNET OF THINGS

SOCIAL MEDIA ACCOUNTS

SOCIAL MEDIA ACCOUNTS OF GIVERNMENT INSTITUTIONS

TV CHANNELS

RADIO

THIS WEATHER-FOLLOWERS' NETWORK IS NOT A CLOSED LOOP. THERE ARE SEVERAL OTHER CONNECTIONS SUCH AS THE ONES CREATED BY THE PUBLIC SOCIAL MEDIA ACCOUNTS OF STATE INSTITUTIONS.

THE CONNECTIONS ARE DYNAMIC – FOR EXAMPLE, NEWLY CREATED CHATBOT ACCOUNTS THAT TWEET HOURLY RAINFALL INFORMATION.

THE DATA CIRCULATED BY THE ONLINE
WEATHER COMMUNITIES ACCOMPANIED A
RISE IN INFOGRAPHICS IN THE LOCAL
NEWS, ESPECIALLY MONSOON RELATED
INFOGRAPHICS.

The rains were late in 2019, which prompted a speculation
about citywide water cuts. An infographic published on
June 24, 2019, showed how both Santacruz and Colaba
monitoring stations had recorded a rainfall deficit as
compared to the same date on other years.

These numbers record the monsoon's
June arrival date in previous years.

SANTA-   COLABA
CRUZ

165.3   127.2
MM   MM

DEFICIT

-197.3   -127.2
MM   MM

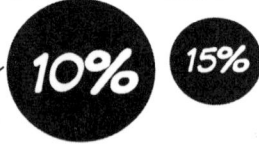

2018...9
2017...12
2016...20
2015...12
2014...15
2013...7
2012...17
2011...5
2010...14
2009...24

These figures
document the
water shortfall.

10%   15%

WATER   2019   2018   2017
DEFICIT - 5.3 + 13.5 +17.6

This statistic shows how the cumulative rain on
June 24, 2019, is much lower than the figures on
the same date from previous years.

By the end of June, the problem had shifted to one of excessive and intense rain. After spells of heavy rain, an infographic showed that this was the second highest 24-hour spell for June in the past decade.

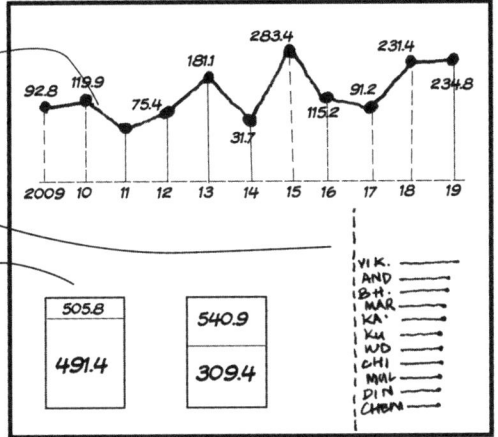

283.4
231.4
181.1
119.9
92.8   75.4   91.2   234.8
31.7   115.2

2009 10   11   12   13   14   15   16   17   18   19

This stat showed the number of areas in the city that recorded more than 200 mm of rain in 24 hours.

Another figure showed how 97% of the total required rainfall for the month had dropped over two days.

505.8   540.9
491.4   309.4

VIK.
AND
BH.
MAR
KA.
KU
WD
CHI
MAL
DIN
CHEN

The infographic explained how a weakening El Niño and cyclonic systems had affected the monsoon.

Rainfall measurement on days with heavy rain in July 2019.

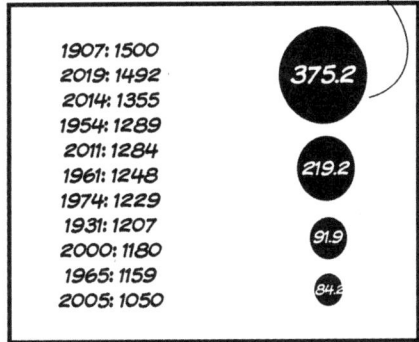

An article on the 30th stated that this was the "wettest July in 112 years." The stats also showed that this trend had emerged over the last two decades.

1907: 1500
2019: 1492
2014: 1355
1954: 1289
2011: 1284
1961: 1248
1974: 1229
1931: 1207
2000: 1180
1965: 1159
2005: 1050

375.2
219.2
91.9
84.2

By September, the monsoon showed no signs of stopping, the rain, at 3,543 mm, broke a 65-year-old record for the highest seasonal rainfall. The previous high was in 1954.

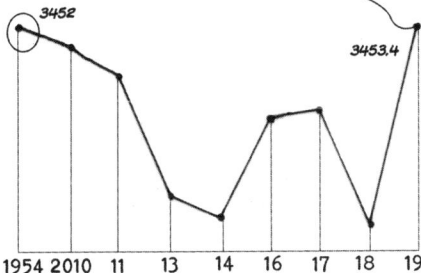

3452
3453.4

1954 2010   11   13   14   16   17   18   19

THE STATISTICAL FIGURES THAT CIRCULATE VIA DIFFERENT MEDIA WORK TOGETHER TO SUSTAIN AN INTEREST IN FOLLOWING THE RAIN AS A MEANS OF NEGOTIATING EVERYDAY LIFE IN INCLEMENT WEATHER. THEY ALSO SHOW HOW THE INTEREST IS SHAPED STRONGLY BY THE DESIRE FOR "FACTUAL" DATA IN THE FORM OF NUMBERS, AND THIS IS EVIDENT IN THE INFOGRAPHICS THAT FOLLOWED THE PROGRESS OF THE 2019 MONSOON.

Infographics, particularly as combinations of statistical images, mediate the anxiety caused by the unpredictability of extremes in important ways because of how they place extremes and patterns in relation to each other. In recent times, as the precipitation pattern changes because of rising global temperatures, the rainfall statistics edge closer and closer to the 2005 figures in the peninsular region (Roxy et al. 2017; Guhathakurta et al. 2011). The downpour of 2005 can be identified as a distinct event in time—a temporal high-water mark—against which later instances of heavy rain are measured. It is also evident in the journalistic chronicles of heavy rainfall such as ". . . another 26/7," and "heaviest rainfall since 2005," where both numbers act as metonyms, indexing presupposed knowledge about the calamitous weather event. The 2005 flood is an event through which the residents of Mumbai reckon time, and each new experience of heavy rain is oriented in relation to it. On the other hand, the graphs that depict newer instances of heavy rainfall also punctuate the experience of monsoon time and its attendant anticipatory anxieties. They create what Kockelman and Bernstein (2012) refer to as a "temporality of metricality"—repeating elements or signs that either set up a pattern or become a way to show something that breaks with a pattern. For instance, in a graph of the daily rainfall in July, the 2005 flood stands out as a singularity. The graphs give numerical and visual solidity to extreme weather because they allow it to stand out against a repeating set of regular rainfall. They also become a way to show how extremes are fast emerging as a pattern.

It has become increasingly common for the MCGM or the IMD's Mumbai office to tweet rainfall information and graphs, which also circulate widely in popular newspapers. This has created a public that is organized around seeing, sharing, and acting in relation to this graphical and numerical information. In recent times, social media, especially Twitter, have become a key means of not just communicating weather predictions and information about phenomena as they happen in real-time, but it is also a medium that shapes users' experience of the event through shared data. Though institutional practices of using social media to communicate meteorological information are relatively new, they have become important ways for the public to access data and to plan their actions around them. For example, during Hurricane Sandy, the number of unique users and tweets jumped substantially as Twitter became an important source of information during power outages. These Twitter exchanges are not just unidirectional, that is, they are not just avenues for concerned citizens to access information but are important nodes of exchange and feedback for institutions that shape governmental response and action as well (Chatfield et al. 2013).

In the early 2000s, extreme weather had just begun to gain global attention. The 2001 Intergovernmental Panel on Climate Change Report contains a section asking whether it is indeed the case that climate extremes are changing, which signals the emergence of a public conversation (Houghton et al. 2001). These emerging conversations clearly influenced the public interest in following the weather, especially events that stood apart from patterns. The works that amateur weather watchers do and the new publics that social media networks create connected to these conversations, and they are also connected to debates on the capacity of urban infrastructure to respond to extreme events. As the social media archive of weather watchers shows, during the monsoon in Mumbai, people are interested in the ways social media combine with local data to give information about rainfall over neighborhoods, flooding in low-lying areas, and real-time traffic information.

In the context of post-2005 Mumbai, infographics that combine weather data, readings from rain gauges, and stats and patterns can be understood as what Susan Leigh Star (2010) terms "boundary objects." These are objects that that allow for "interpretive flexibility," which is to say that they can be read by those with technical expertise and by the public. There may be no consensus in the interpretations derived by both these groups—that is, they may be used toward different ends. Digital weather stations allow for the easy collection of data, which, via amateur weather groups—those who can be seen as sitting between expert knowledge and the public—circulate on the Internet. Consequently, this technical knowledge is rendered locally relevant and used for making day-to-day and hour-to-hour decisions that influence movement in the city.[9] This has ramifications that extend beyond the everyday management of shifting meteorological anxieties: these anxieties are then projected back onto the city's infrastructural systems, garnering public support for redesigning them in relation to the scalar logics of extremes.

Estuaries to Rivers, Surface to Volume

Every year, during the months leading up to the monsoons, the MCGM begins filling potholes, cleaning drains, and performs other infrastructural maintenance works to prepare for the monsoons. A critical part of this monsoon preparedness is the task of desilting the city's drainage channels, which after 2005, has gained much importance. The MCGM floats tenders inviting contractors who bid to clean up particular stretches. According to the terms, the private contractors are responsible for the entire operation, which includes getting rid of the dredged silt and garbage. The desilting works are critical to ensuring that the channels are deep and clear enough for the water to flow. In 2018, I began talking to engineers at the Storm Water Department (SWD) at the MCGM to understand how the drainage system worked, especially in the northern suburbs. Mira S., who was my main contact during this time, is an engineer in charge of overseeing the storm water drains in one ward. She put me in touch with Nilesh, a contractor who had won the bid to desilt a few different stretches of creeks and drains that year. The site where he was working at that time was adjacent to the highway and the channel led directly into mangroves, or an "outfall area" in drainage terms. Irregular, crumbling concrete walls that ended abruptly a few hundred meters before the highway shored up sections of this channel.

Nilesh's job was managerial: he had to hire the JCB (a digger or excavator that is used in construction work, which, in India, is called a "JCB"), the JCB operator, and the pontoon on which the JCB would sit (in order to reach parts of the channel inaccessible from the road), and he also had to ensure that the sludge and waste that was dredged was carted away by other contractors who would provide the trucks and dispose of the waste. Large piles of dark, foul-smelling silt were piled up all along the road leading to the channel. This silt is an amalgamation of sediment deposited by tidal action, raw sewage, and the millions of tons of solid waste dumped every day. His JCB operator, Shaan, was sitting in his cabin inside the vehicle from where he operated the arm in a continuous motion of scooping garbage from the channel and depositing those scoops along the side. Both Nilesh and Shaan were old hands at the desilting business. The desilting work, they both said, was something that the MCGM had been doing since the time before the flood. However, after the flood it had been systematized and followed a proper schedule. I asked them about what happens to the waste that is dredged, and Nilesh described the contracting process in which truck drivers are hired to carry the waste to the dump. As we were talking, Shaan gently placed the JCB's arm on the culvert next to us, emerged from his little cabin, and nimbly climbed over the arm to join us. Nilesh was describing how there was no way to oversee what the other contractors did with the waste and Shaan cut in to say that while their job was to dredge the channel and to ensure that the waste was picked up, that was no guarantee that the waste would be deposited to dumping grounds and not left to fester along the channels. Shaan remarked sarcastically, "I am pretty sure that part of what we dredge is just the previous year's garbage that goes right back."

Desilting is a massive undertaking. According to the figures released by the MCGM in 2018, every year more than forty kilometers of channels and drains must be dredged to remove more than 100,000 metric tons of silt. As Shaan and Nilesh pointed out, there are many places where the JCB cannot reach. In many parts the edge conditions are precarious, which makes their job harder. "Each part is different," Nilesh said, referring to the different parts of the city where he had to work. "For each part," Shaan added, "we have to see how to maneuver the vehicle, whether we need a pontoon to hold the JCB afloat, and where to pile the stuff so it can be easily taken away." From time to time an engineer or staff member from the MCGM would come by to supervise their progress and help them figure out details like access. Mithi, which comprises more than half of the desilting works, is markedly different in each section, requiring as much improvisational thinking as the other channels in the city. The channels have changed over the years as concrete walls have come up along man sections, which they admit has made their job a little easier. However, the amount of waste and silt does not seem to go down.[10]

A little less than eighteen kilometers in length from its origin in Vihar Lake to Mahim Bay, the Mithi runs across what used to be the island of Salsette—which was only included in the city's limits in the 1950s. Until then, what is now referred to as the city of Mumbai (which includes the northern suburbs) was only the southern part, the Island City, which itself was created by reclaiming an archipelago of seven islands. The formal underground storm water drainage systems were installed in the Island City as it emerged as an important port for the colonial empire.[11] The drainage system for the rest of the city—that emerged on the island of Salsette—was created in an ad hoc manner as the suburbs in the north developed (Arunachalam 2005). A plan proposal commissioned by the Mumbai Metropolitan Regional Development Authority (MMRDA) to improve the conditions of the Mithi notes that it originates from the overflow of Vihar and Powai lakes, which are artificial reservoirs built by the colonial government in the early twentieth century (Mumbai Metropolitan Regional Development Authority 2006).[12]

In an article on urban flood resilience, Kapil Gupta (2007, 184) describes the drainage system as "a mix of simple drains and complicated network of rivers, creeks, drains and ponds." This network is made of approximately 330 kilometers of major and minor channels, 151 kilometers of arched drains, more than 500 kilometers of pipe drains, and close to 2,000 kilometers of roadside drains. They work in a hierarchy; the smaller drains empty into the major and minor channels which empty into the creeks and 186 outfall areas, and then finally into the Arabian Sea (BRIMSTOWAD Report 1993). In the reports on Mumbai's stormwater systems published before 2005, the description of the drainage system in the northern part of the city (where most of the flooding occurred) conveys its infrastructural-ecological nature. In a detailed study carried out 1993, which was called the Brihan Mumbai Storm Water Drainage (BRIMSTOWAD) Project (BRIMSTOWAD 1993, 12), the authors write: "The drainage system [of Mumbai] is largely based on *nullahs* (channels/ gutter) which are based on the original natural drainage channels." The authors also state that there is no planned system, no topographical data, and refer to the Mithi as a part of the larger Mahim tidal-creek system where garbage accumulates.

It is only in the reports that follow the 2005 flood that Mithi's identity as a river takes a front seat. The Mithi is no longer described as larger complex hydrological and infrastructural system with a large catchment area. Instead, it is framed as a river polluted by citizens treating it as a gutter.[13] This new focus on rivers is also evident in the subsequent design proposals that were drawn up for restoring the Mithi, which entail desilting and deepening the channel, training its sides, and adding concrete retaining walls. In their new avatar, channels such as the Mithi would have defined edges, clean sides, and deep channels that would carry water out of the city in one direction. Instead of flooding and opening up into large mangrove swamps, the channels would lead to outfall gates that would release this rainwater into the sea while separating the trash that would be carried with it. The reinvention of Mithi as a river is, at its core, infrastructural though couched as an environmental problem brought on by the pressing ecological concerns of climate change.

Central to the politics of river transformation is the concept of "carrying capacity"—a term that is widely used both by planners and ecologists. In the biological sciences it refers to the number of living beings an area can support, and in planning terms, the number of people or the amount of load a given infrastructural system can handle (Zhao et al. 2009). Post-2005, this term was used to both describe the Mithi's degradation (that its carrying capacity had reduced as a result of garbage) and its future transformation (as a river capable of carrying rainfall to prevent flooding). Carrying capacity was critical to engineering the Mithi as a channel that can accommodate larger volumes of water. For that, it could no longer be a part of a meandering, complex system with creeks, wetlands, and estuaries, but had to be remade into a channel based on volumetric capacities.

Carrying capacity is an important measure in the drainage design because it determines the volumetric intensity of water (the amount of water that flows through a system over a given interval) that a system can accommodate. As the National Disaster Management Guidelines ("National Disaster Management Guidelines" 2010) on urban flooding acknowledges, it is not an easy task to fix this number. Its calculation is based on a number of factors such as the total area that the drainage channel serves. It is also determined by the landscape; if there is little development, then the volume of water is reduced, as a considerable amount will percolate through the soil. In urban areas that have a high percentage of paved surfaces, the drains have to not only accommodate a higher volume of water, but also account for the greater speed—called the "run-off coefficient"—with which this water enters the gutters, because smoother concrete surfaces offer less resistance. Carrying capacities are also determined by the amount of rainfall a place experiences, rainfall intensities,[14] and their "return" period—the probability that a rainfall of certain intensity might occur over a given period of time.[15] The city planners have to balance cost and probability to decide the return period they will use to determine drainage design. As the national guidelines state, there is a speculative element to designing drains because it involves predicting how an area might develop. The run-off coefficient may change as more areas become urbanized. The return periods too are "subjective" measures as extreme weather is quickly altering the probability of intense rainfall events ("National Disaster Management Guidelines" 2010).

This uncertainty about measures is evident in the historical design of drainage systems. In most parts of the Island City, with its "formal" underground drainage system, the channels were designed to handle rain falling at the rate of 25 mm per hour. This capacity of the colonial drainage network was calculated in relation to the urban development and weather and landscape conditions from more than a century ago (Gupta 2007). The BRIMSTOWAD Report recommended upgrading the entire city's system to handle 50 mm of rain per hour (mm/h), especially because of the rapid development in the suburbs. A second report proposes to double this figure (Rafiq et al. 2016). While the previous recommendations were calculated on the basis of rainfall data collected from 1843 to 1969, more recent data show that rainfall intensity is rising (Gupta 2007). Moreover, as the first BRIMSTOWAD Report (1993) notes, bringing in a new standard is a hard task given the complexity of the system, its intersection with other systems such as roads, and existing development.

While these measures are clearly shifting as a result of anthropogenic climate change and changing weather patterns, as engineers tasked with redesigning systems describe, it is hard to translate these metrics onto the complexity of existing systems, especially in a dense urban setting. Dr. Kumar (the scientist I quoted at the beginning of this chapter) and his colleagues who were tasked with assessing existing systems and recommending changes were wary of these anxieties and the spectacular shift in the volume of water that new systems would have to be designed to accommodate if they were to cope with another cloudburst.

One morning, as we were walking to a coffee shop near his office, he was talking about cloudbursts and the scale of cumulonimbus clouds that are capable of discharging the kind of volume that descended on the city in 2005. The phenomenon did demand a reassessment of the city's existing systems, Kumar acknowledged, even as he underscored the impossibility of designing systems that could handle the volumes of precipitation associated with cloudbursts. Drainage channels, he said, are usually designed with specific return periods in mind, that is, the average and peak rainfall over periods of five, ten, twenty-five, and fifty years are used to estimate the design and the dimensions of the channels. As Kumar carefully explained, thinking in terms of cloudbursts would render these measures somewhat irrelevant and perhaps even distract from the fundamental problem: that the city needed drainage systems that worked. As he said, "If we have to design for extremes, we will have to think completely differently. Let us forget about [the] cloudburst for a moment, even if Mithi has to accommodate the volume of water that is discharged during heavy rains, it will have to be very different. Because, we all know that it does not take much rain for the city to flood."

In the aftermath of the flood, the MCGM and the MMRDA's main strategy for raising the carrying capacity of the city's drainage was to expand the city's main channel by raising concrete walls and put in place several other measures such as controlling the discharge of sewage and solid waste into the Mithi. These measures were supposed to dramatically increase the volume of water the Mithi could carry and reduce the level of pollutants in it. A Mithi River Development and Preservation Authority (MRDPA) was set up the year after the flood, which was supposed to coordinate the efforts of the MCGM and the MMRDA. In 2017 MRDPA hired consultants to produce a detailed report on Mithi River Rejuvenation Project, which is ongoing.

The emphasis on volumetric capacity, reordering, and channelization is evident in numerous reports and official recommendations published since 2005. For example, a 2006 technical report published by the Central Water and Power Research Station recommended widening and dredging the Mithi to cope with the higher volumes of rainfall in the long term (Central Water and Power Research Station 2006). Similarly, a study commissioned by the Maharashtra Pollution Control Board noted that the carrying capacity of the city's drainage system is severely reduced because of a lack of comprehensive pollution and waste management systems (Klean Environmental Consultants 2006). While these metrics do perform important work in assessing urban systems, given that any infrastructure is not a self-contained object, the ramifications of capacities, design, and metrics extend beyond the systems themselves. The calculations used to manage extreme weather are not abstract measures, but produce deliberate practices, alter the relationship that engineers share about infrastructural systems, and are material measures of political control (von Schnitzler 2017). While carrying capacities offer objective measures, they are by no means easily fixable. Rather, "volume control," as Joshua Cousins (2017) writes, is a technopolitical practice deployed to govern the flow and distribution of resources.

Everyday Engineering

For the many planners and engineers who worked at the MCGM, the 2005 flood was an important marker in time as it ushered several changes in their day-to-day work life and urban systems such as the process of preparing the city for the monsoon, disaster management, and new rainfall infrastructure. It introduced new undercurrents of uncertainty that affected many different aspects of everyday life and decision making in the city, especially in the work life of its engineers. These undercurrents percolated into the institutional cultures of predicting, monitoring, and preparing for monsoons in various ways. They were evident in the facilities and equipment that were installed: because the measuring gauges had ceased to function a few hours after the rain began, resulting in both poor rainfall data on the day of the flood and the failure of critical warning systems, the MCGM installed sixty new automatic gauges at various locations.

These new bits of equipment, systems, and infrastructure demarcated the before-and-after flood time, as did peoples' anxieties about the rain through their personal experience of the flood. That there are clear pre- and postflood times is also evident in the repeated comparisons that cropped up with every subsequent monsoon, where heavy rains are invariably measured against the 2005 floods, and then, so is the MCGM's performance in keeping the city running. The municipal engineers and contract workers who maintained and upgraded the drainage systems of the city would often remark about the changes that took place after the flood and the physical traces that remained in the form of water lines on walls, dumped and damaged vehicles that remained by waysides, and broken infrastructure that took a long time to fix. One of my contacts at the MCGM once told me that he remembered how, even some years after, he could see the height to where the water had risen on the damaged walls. The improbable statistic had left a physical trace on the city and a definite mark on the relationship that the engineers shared with the infrastructural landscape.

During my visits to the SWD at MCGM, I usually made a beeline for Mira's desk, an engineer who oversees drainage systems in one of Mumbai's municipal wards. Mira had agreed to allow me to go through the city's master plan document and other reports at her desk. I was not allowed to take the reports away from the desk, which restricted my movement but came with other advantages: I was able to ask Mira questions as I worked my way through the documents to gain a rudimentary understanding of this complex system. Mira liked to say that the SWD "has one main job –to make sure that the water in the drains keeps moving," a line she would often repeat to me. It was when the water stopped moving that problems arose, but while this moment of dysfunction made infrastructure visible, it did not necessarily render it transparent to everyone:

> As long as the water stays away, people are fine. The minute the water begins to accumulate, then there is a worry that the system is not working. But what exactly is the system? That no one wants to ask because

no one has the time to understand. You need an engineering degree to understand how it works and all the technical details . . . and the [infrastructural/ drainage] condition in Mumbai is such that technical knowledge is not enough, you have to learn many other things—where there are choke points, or some other details about that site—to get things to work.

Mira's desk was a great vantage point for getting a sense of how engineers worked with the complexity of systems. Mira's juniors would head for her desk when they wanted to figure something out. Mira would then sit with the plan and try to figure out solutions to that problem. Each problem required a different solution based on site-specific conditions:

> So if we look at this problem [a colleague had come to her with a question about constructing a drainage line from a site to the nearest main channel], we can see that the lines run here and here [gesturing to the lines located on the plan drawing], but we need to figure out how to connect the drainage from the construction site to it [one of the main lines]. We can use the [construction] plan, but we also need to go to the site to figure out how to make the connections. On this site, the road on one side is narrow and on the other side, we have slums. There is an existing line, which we can use if we can figure out the junctions. A lot of times, these kinds of problems are figured out on site.

The plan drawings, Mira said, conveyed a diagrammatic sense of the system and that was their function. However, one needed a lot more information than the diagram. The diagram could be used to formulate an initial strategy, but much had to be figured out based on site conditions. It was not just engineers who performed this improvisational thinking, but also contractors who were charged with desilting, and private builders who were allowed to construct their own drains provided they followed the MCGM's guidelines.

As anthropologists studying engineering practices and urban infrastructure have pointed out, tinkering and improvisation are key elements of what engineers do with urban systems in the course of their day-to-day work (Björkman and Harris 2018). This is especially the case in a city like Mumbai where the infrastructural systems are often old and stretched well beyond the limits of their design, and thus, working with these systems entails a constant negotiation. This improvisation is not limited to the physical infrastructure, but also to knowledge of that infrastructural system and the other institutions, enterprises, and people it connects. The MCGM often transfers its engineers between different departments. Consequently, Mira and her colleagues have to learn the working mechanisms of their respective departments quickly and rely on the informational flows from their predecessors and seniors, a function that Mira performs for her junior colleagues. However, Mira and her colleagues would often point out to me that given the needs of a massive city with millions of people, we needed reliable systems that meet specific standards and as such, the current systems are in dire need of repair and upgrades. Repairs and adjustments that happened on a smaller scale were insufficient in addressing the needs created by the changing weather patterns.

Scholars of urban flooding and disaster planning have written about how events such as floods are in fact planned events that are outcomes of planning and development policies as much as they are of freak weather events (Bremner 2020; Goh 2019). Municipal engineers and workers are keenly aware of this and are constantly tinkering and working with failing systems and policies. Mira and her other colleagues would ruefully say that when the city floods, they too have to wade through the waters. Almost every conversation about storm water systems or the monsoons is shot through with the uncertainty created by the 2005 flood. A sense of a postflood time and postflood ways of working in the city also marks their personal experience of the event. These experiences reshape not just how engineers work with infrastructure, but also alter ideas of scale and the ways in which they come together with infrastructural systems to mediate the needs of a city. They manifest as the desire to find infrastructural solutions that become the material and symbolic manifestations of the anticipated future (Groves 2017).

The city's channels lead to outfall areas, and these are often mangroves and mudflats—soft, marshy areas that soak up the rain and tidal water. Gates and pumps control some of these outfall areas and regulate the water levels in the drains regardless of the state of the tide. After the 2005 flood, the MCGM proposed adding a number of new pumping stations that would keep the tidal water at bay while pushing out the rainwater. Many of these stations were proposed on wetlands and to get around the environmental clearance problems, the MCGM invoked the Disaster Management Act (2005), which allowed it to exercise eminent domain.

Pumping stations are massive and complex machinery that are installed at the end of major channels, at the point where they meet the sea and they sit like dividers separating different waters. Mira had managed to get us an appointment with the engineer, Purva, at the station. Purva took us through a tour of the facility, one part of which was a massive set of gates that separated the main drainage channel from the sea. Both sides of the pumping station—the cityward and the seaward sides—were occupied by a slum, which housed informal industries. Purva pointed to one side of the pump where the garbage was separated from the water before it went out to the sea. Mumbai's lack of a working solid waste management system is one of the reasons why its drains do not work and stations such as this one, Purva explained, performed the dual functions of regulating the water level in the channels and clearing up blockages.

Purva described a thrill in seeing something massive like this at work—a thrill that Mira also shared through her animated explanations about infrastructural systems. Yet, as Mira said over tea one day, these systems required a completely different scale of thinking and a different way of working with the machines, structures, systems, people, and the city. These new and proposed infrastructural systems sever the connection between mangrove landscapes and the channels that convey storm water, the mixing of which is important for sustaining the brackish landscapes of the outfall areas.

Attending to the mixture of urban waste and mud, pumping stations that mediate the connection between salt and fresh water, and how concrete articulates channel edges is a means to think more broadly about how urban environments are socially and technologically constituted. The transformation of the Mithi necessitated systems that would allow greater control over the separation of rain and tidal water, while also rescaling these systems in relation to new volumetric sensibilities. Simultaneously, this reimagination of the Mithi as a different kind of infrastructural and ecological assemblage also gave rise to different spatial politics and claims. Volume became an empirical measure for extracting territory (Elden 2013), and its accompanying changes a channel for retraining capital (Bakker 2010).

River Edge and River Lives

To meet new carrying capacities the MCGM and the MMRDA began widening the Mithi and shoring up its edges with retaining walls. This process became a means for displacing the informal communities that lined these channels and creating saleable land and green eco-zones. These proposed riverine landscapes are "fantasy plans" (Weinstein et al. 2019) that do very little to address the problem of climate-related flooding. Yet, these plans were met with very little protest, and instead, a new movement has emerged in recent times where groups of middle- and upper-middle-class citizens organize "river marches."[16] The River March collective has garnered the support of regional right-wing parties such as the Shiva Sena and the Bharatiya Janata Party. Their environmental vision is rife with nationalistic rhetoric and the movement scapegoats informal settlements as the source of the city's drainage problem.

The River March collective began with a small group of people coming together to clean the channels, quickly swelling as residents were drawn to its "bourgeois environmentalism" (Baviskar 2020). As Sapana Doshi (2013b) describes in her meticulous account of the displacement resulting from Mithi's restoration, despite evidence to the contrary, numerous reports named slums as the source of trash and sewage that choked the channels. The movement glommed onto this narrative, which also found support from politicians who called for the immediate demolition of slums and supported river-edge development. By late 2017, the marches had turned into large public events led by politicians. In their public addresses, the politicians would urge the crowd to "remember" and "take-back" their rivers. The movement was not without its controversies—in 2018, the then chief minister's office was found to have used public funds to produce a music video titled *Mumbai River Anthem*, describing the "four rivers" as the pride and lifeline of Mumbai in years past. These evocations of lost rivers and pleas to "remember" them naturalize a geographical memory of a riverine Mumbai and lay the foundation for large-scale infrastructural programs. Instead of a complex channel, practices of remembrance borne from a metrics of capacities reconceive rivers in ways that negate plurality (Ley 2018). They have acute effects for those who rely on the river (such as salt-pan workers, fisher communities, and cattle owners) as they are often framed as figures that pollute the river.

In Mumbai, the creation of a historical narrative that combines natural and sociocultural landscapes (Rademacher 2011) brings together the postdisaster rhetoric of environmental responsibility and disaster preparedness programs that anticipate future events—in this case, preparing the storm water system to carry volumes of water generated by intense monsoon rain. By categorizing the Mithi as a "river" and focusing on volumetric logic of capacities, the state, along with developers and higher-income citizen groups, were able to impose a new order on the city's hydrological landscape. The consequences of Mithi's changing edge conditions, displaced informal settlements, reclaimed marshes, destroying expansive mangrove systems, capturing land for real estate development (Parthasarathy 2015), further entrenched the politics of belonging that divides Mumbai. Recently, when Aaditya Thackeray, a Shiva Sena leader who is the current Chief Minister Uddhav Thackeray's son, became the environment minister, he announced that the Mithi River restoration would be the ministry's main focus (*Mumbai Live* 2020). The changing Mithi River is but one chapter in a longer history of how right-wing regional parties like the Shiva Sena deploy technology, identity, and belonging as a means of staking claim to the city.

While the newly emerging river has altered the city's landscape in radical ways, Mithi does have other modes of existence. The older reports on the city's drainage testify to this—they note the ways in which communities would constantly work with these channels, such as damming them to irrigate agricultural land.

They capture the ways in which these channels are dynamic and intersect with urban communities in complex ways to create "amphibious" ways of being (Ten Bos 2009). To read these accounts and to find these relationships is to recall a different kind of river that muddied the state's problematic ecological visions, to acknowledge the fundamentally porous nature of Mumbai, and to find ways of living with water that are equitable, instead of doubling down on walls, reclamation, and solidity.

IN 2017, I MET WITH ASLAM SAIYAD, A PHOTOGRAPHER WHO WAS DOCUMENTING THE URBAN LANDSCAPE OF THE CITY'S CHANNELS. OVER THE PHONE HE HAD SAID THAT HE HAD BEEN CALLED TO DOCUMENT ILLEGAL DUMPING IN A WESTERN SUBURB.

WE MET AT THE SANJAY GANDHI NATIONAL PARK, A SHORT BIKE RIDE AWAY FROM THE SITE, WHICH ADJOINED THE DAHISAR RIVER. THERE, WE WERE MET BY A MAN WHO HAD CALLED ASLAM THERE AND WAS A RESIDENT OF A HIGH-RISE HOUSING COMPLEX.

THE THREE OF US PEERED OVER THE COMPOUND WALL, AND THE RESIDENT POINTED TO SLUDGE FROM AN OLD STABLE BELOW THE BUILDING.

WE CROSSED THE STABLE, LEAVING ITS WARM SMELLS COWS, DUNG, AND MUD BEHIND, AND WE WALKED OVER TO THE OTHER SIDE, INTO A SMALL SETTLEMENT WHERE WE ENCOUNTERED A LONG ROW OF POTS AND BUCKETS IN THE MAIN LANE, PLACED IN LINE FOR WATER FROM A COMMUNITY TAP. FROM ACROSS THE CHANNEL, WE COULD SEE THE OUTLET FROM THE STABLE THROUGH WHICH A SLUDGE OF MANURE MIXED WITH HAY HAD ACCUMULATED ON THE DRY BED, WHICH WAS PART ALGAE AND PART TRASH.

WE WALKED ACROSS THE HIGHWAY AND DESCENDED TO THE RIVERBED. PEOPLE WERE WASHING CLOTHES IN THE SMALL STREAM THERE, AND NEARBY, WORKERS WERE TAKING A BREAK FROM CONSTRUCTING A CONCRETE WALL TO SHORE UP THE NEWLY WIDENED CHANNEL.

LOOKING AT THE MASSIVE CONCRETE WALL, ASLAM WRYLY REMARKED THAT THE STABLE HAD EXISTED LONG BEFORE THE HIGH-RISE; IT HAD LONG USED THE DAHISAR TO DISCHARGE ITS WASTE AND WASH ITS COWS, IT WAS CONVENIENT TO SAY THAT SLUMS AND STABLES POLLUTED THE RIVERS AS IT TOOK THE ATTENTION AWAY FROM THE REAL PROBLEM OF UNPLANNED DEVELOPMENT AND MADE IT EASY TO JUSTIFY URBAN DISPLACEMENT.

LATER, WHEN I WAS OUT WATCHING THE DESILTING HAPPEN, OR WHEN SALIL AND I VISITED THE PUMPING STATION, I WOULD THINK ABOUT ASLAM'S WORDS AND HIS PHOTOGRAPHS, WHICH EXPLORED THE CITY AS HE MADE HIS WAY DEEP INTO THE NATIONAL PARK.

HIS PICTURES SHOWED A DIFFERENT IDEA OF A RIVER THAT WAS NOT ABOUT THE SEPARATION OF MATTER – DIRTY FROM CLEAN WATER, RAIN FROM SEA, AND SILT FROM THE BRACKISH.

INSTEAD, PEOPLE, ANIMALS, AND PLANTS MINGLED. THE WATER SEEPED INTO THE LANDSCAPE. LOVERS SAT BY THE EDGE. CHILDREN WADED AND DOVE. ANIMALS WANDERED AND PLANTS SPRANG FROM THE MUD.

On statistical graphs of rainfall histories, trends, and averages, the 2005 flood stands apart from the pattern as an anomalous blip conveying the hitherto unimaginable volume of precipitation. These graphs are a part of a numerical consciousness—of apprehending experience through numbers (Guyer et al. 2010) and a means of reckoning with extreme weather phenomena such as the increasing incidence of intense, high-volume rainfall and with the increasing incidence of urban flooding. These shifting scientific calculi of sustainability in turn impel political consensus on planning and urban interventions.

These blips on the graphs that mark the incidences of heavy rainfall translate into the urban landscape in different ways. They result in an anomalous landscape designed for combating extremes, where seawalls keep out rising waters, massive outfall gates pump out tidal inflow, and where the city's hydrological terrain is completely reimagined as a system with the increased volumetric capacity to handle intense rainfall and prepare for future floods. "Preparedness," Lakoff (2017, 14) writes, defines a "generally agreed upon terrain for the management of threats to collective life." These infrastructural projects are material manifestations of preparedness efforts and that mitigate future risks in the present. Statistical images are mechanisms for assessing those risks and defining the scale of infrastructural interventions, and thus, delineating the boundaries and frameworks of disaster management.

River restoration is a means to capture land and reposition cities in a changing global economic order (Coelho and Raman 2010). The quest to make cities sustainable has not done away with the rising asymmetries in the distribution of rights, asymmetric access to public amenities, and increased precarity for poorer urban residents and the city's extra-human worlds. These infrastructural changes that are remaking the coast are also remaking urban politics. The large-scale interventions and policy changes introduce material and economic risks for Mumbai's coastal communities who have begun mobilizing and creating collectives and finding new channels to gain a foothold in the planning process.

4

UNFOLDING PLANS

Drawing, Time, and Spatial Politics

Scenes from a Seafood Festival

Versova Koliwada is located on the edge of Andheri West—one of Mumbai's affluent suburbs. The Malad Creek that runs along one side of it is part of an intertidal zone with dense mangroves, in the middle of which is a large sewage treatment area that dumps treated waste into the creek. Depending on the tide, the water in the creek moves back and forth swirling all the matter it collects, both treated and untreated, and eventually dumping everything out toward the sea. While the marshy edge of the creek and the beach along the sea provides a natural and obvious boundary, the entrance to the settlement on its landward side is marked by a traffic island with a statue of a fisherman and a woman with a boat. Here, the road makes a sharp right turn, and this statue designates the area beyond the stream of traffic as a koliwada.

The edge is more blurred than it appears. From the point where the main road turns and runs toward the settlement area, it is lined with homes that overlook the beach. The residences are of different sorts. Between the beach and the residences are spaces where fish is dried, catch is sorted, and the fishing gear is stored. Parts of the Versova fishing village extend beyond the tight residential core and are closely interwoven into the surrounding urban fabric. A comparison of a historical map of Versova Koliwada with a recent satellite map hints at the growth of both the surrounding areas and the settlement that has happened in the last few decades.[1]

The Versova Seafood Festival usually runs for three days in late January and is attended by hundreds of thousands of people annually. Versova Koliwada was the first to hold a seafood festival in the city and, in the past few years, several other fishing villages have adopted this program of having an annual festival, which serves as an important public forum to gain support for the Koli community's land claims. When I attended the festival in 2016, by eight o'clock the festival ground was filled with thousands of people. I was there with Sagar Shiriskar, a photographer who was collaborating with me on documenting the festival. On the wall behind the podium and backstage area were large posters sponsored by the Shiva Sena, and the Maharashtra Navnirman Sena. Bal Thackeray, Raj Thackeray, and Udhhav Thackery's visages loomed over the audience, overlooking the festival space. The food stalls had canopies made of pristine white fishnet. Caught in this net were polystyrene replicas of pomfret, *surmai, bombil,* lobster, and shrimp—the most sought-after catch. The stall that I was standing under had a picture of a Koli woman in a blue sari and a mechanized boat. The stall to the right had a picture of a Koli couple, and to the left was an advertisement with pictures of children in traditional attire that had perhaps been cut out of old studio portraits. At the festival, men and women dressed in traditional clothes cooked in the kitchen, manned the stalls, bussed tables, and guided the crowd to their family's stalls. Occasionally one would be asked to pose for a picture, and the crowd would part. They would then be photographed against a background of the stalls, fish, and pictures of other Koli men and women.

Mr. Ajay, one of the organizers of the festival spotted me and came over. Like the other organizers and main guests, he was wearing a business suit with a tie, and a red fisherman's cap on his head. A roar erupted through the crowd; on the podium, the band had struck the first notes of "Mi Hay Koli" (I am Koli), a very popular number. Onstage, the dancing had ended, and several guests had been ushered to the microphone to address the crowd. A man dressed much like Mr. Ajay in a business suit and a red Koli cap came on stage and addressed the crowd with the words, "Mi hay Koli!" The crowd echoed his words back. Then, assuming a more serious tone he said,

Following the 2011 Coastal Regulatory Zone (CRZ) notification, one of the most critical points in the Koli community's quest for housing and land rights was their ability to prove that the lands they occupied were historic fishing villages. It was not enough to prove their identity as native inhabitants and the traditional fishing community of the city; they also had to prove the identity and historicity of the land they inhabited. In the absence of legal documents that could be submitted to planning authorities, the proof of these claims lay in constructing visual evidence that marked the space and the community as distinctly koliwada and Koli respectively. It was essential that these be visual signs be immediately recognizable by other citizens of Mumbai, and especially, to any government surveyor who might visit the koliwada to ascertain its identity for inclusion in the Coastal Zone Management Plans (CZMPs). As my conversation with Mr. Ajay reveals, much hinged on recognition: in popular events such as the seafood festivals, it fell on the visual signs that circulated within the festival grounds to express the villages' identity and that of its residents.

This chapter looks at how visual work is an integral part of indigenous identity and land claims. A key figure is that of the traditional Koli fisher—particularly, a Koli woman—and the community operationalizes it in tandem with other images, such as surveys. Thus, indigeneity, belonging, and history emerge not via utterances of "mi hay Koli" alone, but through a plethora of visual signs that come together to give weight to those identities, claims, and narratives.

Both in popular imagination and academic works, the Kolis, Mumbai's indigenous fisher community, are depicted as people who witness the birth of the city (Punekar 1959). For instance, S. M. Edwardes (1912), British civil servant posted to the Bombay Presidency, describes Kolis as stereotypical residents of the colonial port town. Edwardes's account includes illustrations by M. V. Dhurandhar, an artist who became the first Indian director of the Sir JJ School of Arts, Mumbai's first art institution. Dhurandhar painted the city's residents as singular figures set against the blank backdrop of a studio, illustrating the "types" of people who inhabited the city. The book has an illustration of a Koli man, identifiable by his cap and by the oar in his hand. On another page, a Koli woman is heading to the market—she is identifiable by the drape of her saree and the basket of fish on her head. Dhurandhar's images keep with other colonial depictions of native subjects through the use of signs that mark their caste identity (Pinney 1997).

The Koli community's history is one that is marked by systematic subjugation. When the British acquired the islands of Bombay, Kolis formed the predominant native population on the island and "the essential basis for tax and rental income" (Reeves et al. 1996). The British exploited the Kolis by setting up new forms of fisheries control and relegating them as "coolies" performing low-wage labor (Reeves et al. 1996). Paradoxically, the Koli community's identity as being indigenous to the city provided it a measure of protection against gentrifica-tion programs. In recent times, this indigenous identity has become central to their political negotiations with planning authorities, developers, and the city's elite (Chhabria 2018). The CRZ changed this situation somewhat—it did not question whether the city's fishers were Kolis or if they were indeed indigenous. Instead, it questioned the status of the land they resided in. From the very start, the question was posed as a visual one as it concerned identifying and map-ping the extents of historic villages. Thus, the response from the Koli community too was visual: to create maps and plans for their settlements and to make their identity—and by extension the identity of their villages—evident.

The chapter focuses on two examples: the annual seafood festivals that are held in the larger fishing villages (koliwadas) and participatory planning projects that took place in two villages between 2011 and 2012. The visual work at these sites includes surveys and cartographic plans, and elements drawn from older image-making practices, such as Dhurandhar's studio portraits. The examples work in different ways—the plans and surveys, which mirror the visual struc-ture and language of official plans, make spatially grounded political claims and demonstrate the Koli community's ability to develop their villages. The visual spectacle at seafood festivals recalls their indigeneity for a city and state that has forgotten it. At the seafood festivals, the community draws on long histories of constructing racial difference through the visual (Poole 1997). The two examples demonstrate how the visual and political fields collapse onto each other (Kaur 2003). Maps, photographs, diagrams, plans, Styrofoam fish, and nets are material objects that act together to construct the Koli community's political narrative. As Chris Pinney (1997, 8) writes, images are not just "a mirror of conclusions estab-lished elsewhere" or "illustrations of some other force"—they are where politics happen, "an experimental zone where new possibilities and identities are forged." While plans worked well to construct spatially grounded claims, the visual work at the seafood festival helps ground these claims in history by confirming the community's indigenous identity and the identity of their settlements as fishing villages.

These visual objects articulate indigeneity in specific ways (Li 2000). I take the idea of articulation from Tania Murray Li who writes that it would be simplistic to think of a community's identification as indigenous as "natural" (as the CRZ policy does). Rather it is a

> positioning which draws upon historically sedimented practices, land-scapes, and repertoires of meaning, and emerges through particular patterns of engagement and struggle. The conjunctures at which (some) people come to identify themselves as indigenous, realigning the ways

they connect to the nation, the government, and their own, unique tribal place, are the contingent products of agency and the cultural and political work of articulation. (Li 2000, 151)

The new coastal policy, histories of planning, and urban activism are key elements that have shaped the Koli community's politics in recent times (Kamath and Dubey 2020). Thus, they are also important elements in the Koli community's visual articulation of their identities.

A Time to Draw

When the new 2011 CRZ policy was released, koliwada residents began holding meetings to discuss its threats and opportunities. The biggest danger appeared to be that under this new policy, it was now possible to effect slum rehabilitation schemes along the coastline of Mumbai, which the previous version of the policy did not allow. The previous version of this policy, the 1991 CRZ policy, deemed coasts as ecologically sensitive zones and imposed strict development regulations. This meant that the residents of informal settlements along Mumbai's coast could not expect to be moved into permanent housing under the 1995 Slum Rehabilitation Act (SRA). At the same time, since coastal areas commanded high prices in the real estate market, residents in both fishing villages and slums were under pressure to sell their houses and land to developers.

The 2011 CRZ changed the situation as it cleared the coast for development and made provisions for resettling informal settlements. The policy deemed that "traditional" fisher communities could have a say in how their settlement developed. Residents of informal settlements identified as "slums" would be rehoused under the 1995 SRA. Since most fishing villages (called koliwadas) are part of larger informal settlements, the policy created different sets of rules for communities that otherwise live in the same settlement. Moreover, since there were (at that time) no existing land records that verified the location and boundaries of all the koliwadas in Mumbai, the residents feared that they would be categorized as slums. The residents of the fishing villages would then be allotted apartments in accordance with the SRA, regardless of the size of their houses. They would also forfeit the profit from development, access to commons, and would lose the spaces and facilities needed to support the fishing industry. Given the high value of sea-facing real estate in Mumbai, the fisher community stood to gain a lot if they were able to retain the right to develop their land.

The second development that lent urgency to this issue was the Municipal Corporation of Greater Mumbai's (MCGM) decision to revise the city's

Development Plan (DP). As the plan is a legal document, the Kolis feared displacement if the revised DP labeled their neighborhoods as slums. Simultaneously, the DP's revision was an opportunity to engage the planning process and, since documents that recorded the locations and extent of fishing villages were hard to find, to gain official recognition on a legal document. Time was of essence: the DP controls all development in the city, and the fishers thought it important to intervene and stake their claim before it was finalized and released.

The fishers were not the only community interested in getting involved in the DP's revision. The MCGM had decided that the new DP would be the result of a participatory process and as a result, several planning initiatives sprang up in the city. The most important of these was spearheaded by an NGO called the Urban Development Research Institute (UDRI). The UDRI collected public documents and urged the city's residents to submit their suggestions and objections to the MCGM. Consequently, the Koli community's participatory efforts were a part of a constellation of unofficial planning enterprises taking place at the same time.

The fisher community's immediate goals were to ensure that they were not displaced and to retain development rights to their settlements. The village representatives I spoke to framed their activism as the means to develop their settlements on their own terms—as opposed to succumbing to pressure from builders who sought to take over their land. There were other long-term issues at stake—namely, the tension between the coast as an ecological entity and the coast as sea-facing land, which has a high value in the real estate market. If the coast was given over to the forces of the real estate market, it would result in gentrification and the loss of vibrant coastal ecologies and economies that supported nearshore fishing. Yet, the immediate fear of displacement made it hard to focus on the long-term issues.

The 2011 CRZ policy was released around the time that the DP revision began. Hence, both the Coastal Zone Management Plans (CZPMs—cartographic plans that are vital for implementing the CRZ policy) and the new DP were underway at the same time. The fact that both drawings were in the works—in different stages of survey, drawing, or drafts—presented an opportunity to intervene in the urban planning and the coastal planning process. Also, given that the policy and the development plan required the creation of drawings, it makes sense that the Koli acted visually—by drawing community plans with the help of a local academic institution, and through the visual spectacle of the seafood festivals. For example, the women's wing of the Maharashtra Machchimar Kruti Samiti (MMKS) conducted a mapping exercise to locate all formal and informal fish markets in the city, in time for the DP's revision in 2014. During the Consultative Workshop on the Development Plan (2014-34) through a Gender Inclusive Lens, representatives from MMKS had pointed out the need to safeguard and develop marketplaces, which are sites where Koli women play an active role. Subsequently, the MCGM, MMKS, and the International Collective in Support of Fishworkers, which is a global organization, embarked on the mapping exercise. It would be too simplistic to characterize these sorts of actions as responses to state-led drawings; they also stem from rich histories of design pedagogy and mobilizing communities through participatory design. These histories imbued the Koli community's visual projects with political force, even if these projects lacked official recognition—that is, despite their informality or unofficialness.

BOTH THE DP'S REVISION AND THE CRZ POLICY SPURRED A NUMBER OF COMMUNITY AWARENESS CAMPAIGNS.

IN EARLY 2012, AN NGO WAS RUNNING A DEVELOPMENT PLAN AWARENESS CAMPAIGN IN AN INFORMAL SETTLEMENT.

THE CAMPAIGN WAS ORGANIZED AS A STREET PLAY: THE ACTORS WOULD DANCE AROUND IN A CIRCLE, ALL THE WHILE SINGING THAT THEIR FRIEND "VIKAS" WAS COMING, AND THE RESIDENTS OUGHT TO LISTEN TO WHAT VIKAS HAD TO SAY.

AAYA RE AAYA LOGON RE LOGON

VIKAS KI BAATEIN SUN LO SUN LO

VIKAS, IN HINDI, MEANS "DEVELOPMENT," AND IT IS ALSO A GIVEN NAME.

VIKAS, WOULD THEN MAKE HIS WAY TO THE STAGE WEARING SIGNBOARD WITH THE DP ON ONE SIDE, AND SCREENSHOT FROM GOOGLE EARTH ON THE OTHER.

HERE'S MY LANE, HERE'S THE TREE, AND HERE IS MY HOUSE!

ONE BY ONE, HE WOULD APPROACH HIS CAST MATES, ASKING THEM TO FIND THEIR HOMES, FIRST ON THE GOOGLE EARTH IMAGE.

THE ACTOR WOULD HAVE NO PROBLEM.

BUT THEN VIKAS WOULD TURN AROUND AND ASK THE SAME ACTOR TO FIND THEIR HOUSE IN THE DP, AND SINCE INFORMAL SETTLEMENTS ARE NOT MAPPED IN THE DRAWING, THE ACTOR WOULD FAIL TO FIND THE HOUSE.

VIKAS WOULD THEN TURN AND EXCLAIM THAT SINCE THE HOUSE DID NOT EXIST IN THE DRAWING, IT AND THE RESIDENT DID NOT EXIST IN THE EYES OF THE STATE, WITH THE CAST REPEATING THIS IN CHORUS TO ITS AUDIENCE.

THE PLAY ENDED WITH THE MESSAGE THAT AS THE MCGM WAS REVISING THE PLAN, THE TIME TO ACT WAS NOW, WHILE THE DRAWING PROCESS WAS STILL UNDERWAY. AND THE WAY TO ACT WAS TO SEIZE THE MOMENT OF REVISION AND TO THINK ABOUT WHAT THEIR PLAN FOR THE SETTLEMENT COULD BE.

Image-Document

In August 2010, a batch of first-year students at the Kamla Raheja Vidyanidhi Institute for Architecture (KRVIA), a college in the western suburbs, went to Moragaon Koliwada to study the built forms that made up the low rise, high-density urban settlement. The students were immediately questioned and promptly told to leave after the community expressed their displeasure of measuring and drawing—activities that are associated with the government's demolition drives. However, a little while later, Mr. Rupesh, who introduced himself as a village resident, approached the professors at the school with a proposal to make a development plan for the village. In common parlance, a development plan refers to any cartographic plan document that controls the transformation of an area. In Mumbai, the term (when capitalized) Development Plan (DP) is used for the city's official plans that determine its growth over two decades.

The idea was that this plan would provide possible solutions to the problems faced by the community: ways to increase housing and add civic infrastructure such as roads, garbage collection, and additional toilets—facilities that the community lacked. Mr. Rupesh thought the DP would block any attempts to seize village land while making the most of the development rights the CRZ conferred to the fisher community. The architectural college agreed to this proposal, with the understanding that the plan that the students would produce would only be "conceptual"—a plan that presented *ideas for possible solutions*, and not necessarily an implementable plan. As the conversation progressed, the faculty decided that the project could be a part of their fourth-year urban design studio, which senior students take in their penultimate year of the undergraduate studies. The studio exercise added another koliwada, Malvani, and this village became my primary field site.[2] Soon after, the studio was expanded as a joint exercise with the students from Tata Institute of Social Sciences (TISS). Right from the start, both schools emphasized that this was an academic exercise, and thus, the plans would have a limited scope in terms of execution, and that any official or executable plan was beyond the studio's scope.

Since the villages were located quite far apart from each other, the class split, with each group taking one koliwada as its design site. While Moragaon was quite close to the school, Malvani was located further to the north in a suburb called Malad. During our preliminary visits to the koliwadas, it was also clear that both the settlements were quite dissimilar. The two koliwadas also had a different history and entirely different socioeconomic make up; Malvani had more people directly engaged in the fishing industry as compared to Moragaon. Each village faced different tensions and land pressures. In Moragaon, the municipality had conducted demolition drives at different times. Many of the older residents of the village who lived in considerably large houses faced the unwelcome prospect of having to relocate to a substantially smaller apartment in rehabilitation schemes. In Malvani, the threat was the piecemeal disappearance of village land. Many landowners were selling parcels surrounding their main residential area to private developers to build high-rise apartment houses.

The institutions decided that the studio would run from July 2011 to April 2012, across two semesters.[3] The first few months would be spent in data collection, which meant gathering surveys, government documents, and historical information that would provide the students with base information for their plan proposals. The first phase would culminate in a participatory planning exercise during a "Winter Workshop," where the students would spend ten days at their respective sites in conversation with residents. From both KRIVIA and TISS's point of view, a protracted engagement between the students and the community was necessary because the main goal of the exercise was to teach the students the value of involving the community in the planning process. While the colleges preferred an open-ended engagement that would result in a rigorous study and conceptual proposals in the first phase, the fishers (especially the elected village representatives) were keen on producing a presentable development plan. This was clear from the many notices announcing surveys and mapping exercises during the different stages of the studio. Each of these notices announced that the college students were there to make a *vikas arakhada* (development plan) for the koliwada.

In the first stage, the students conducted surveys in the villages and compiled a preliminary report with cartographic drawings that showed the community makeup, history, building heights, building use, and existing amenities and road networks within the village. In Malvani, the students and faculty would meet with the community members in an open shed in the middle of the fishing village, an event space that doubled as a drying and net mending area during the day. The first few meetings were well attended. However, as the studio wore on, the crowds thinned. It became clear during the very first surveys that there were deep divides between different sections of the village—fishers and nonfishers, Kolis and non-Kolis, Hindu residents and Christian residents, and tenants versus landowners. Not only were a number of residents completely unaware of the project, but others also questioned the fishers' right to commission a plan when the fishers did not necessarily represent the interests of residents who either did not fish or did not belong to the Koli caste. For instance, one of the residents interviewed by the students said, "They [the fishing community] might want to do something for the village, but they have not told us anything. Now you are coming and asking me to list the problems I have. Why should I talk to you?"

Malvani's residents had different stakes in any potential redevelopment project because of the complex tenure systems in the village. Even within the Koli community, each house typically contains multiple households occupied by various family members with a claim to a share of the ancestral home. Many family members or tenants did not fish and often had jobs in other parts of the city. Several landowners in the village had built additional residences, which they leased to migrant workers looking for cheap housing. Though the students did interview members from different communities in the village, these groups never took part in any of the public meetings and expressed their disapproval and suspicion of the participatory process initiated by the members of the fisher cooperative. By the time the Winter Workshop started, the only people attending the public meetings were either the members of the cooperative society in the village, their relatives, or members of fisher collectives from other villages.

At the meeting held in the cooperative society's office space, the faculty and students from both schools voiced their concerns regarding the lack of participation from different community members. Consequently, they tried introducing new strategies to increase participation and produce a plan, that as one faculty member said, "actually addresses their [residents'] needs, rather than imposing something that we might think of as a lack." This was one of the goals of the Winter Workshop: to get the students to interact with as many residents as possible not just to make a viable plan, but also to ensure that the pedagogical goals of the exercise were met, which was to teach them how to arrive at more just and equitable urban outcomes through consensus-based planning. By this time, members of the fisher cooperatives and other local representatives were anxious that the studio had gone on for a couple of months without yielding any plans. They requested that the colleges provide them the final plans, and they repeatedly asked for these drawings during various meetings. On TISS and KRVIAs' part, such a plan could only be drawn after a rigorous study had been undertaken, as the contents of such a plan would be critical.

The disparity between what the plan meant to the colleges versus the meanings and values assigned to it fishing community came to a head during the Winter Workshop. The fishers thought it was important that the plan follow the visual format of the official DP so that it could be shown to surveyors, developers, and civic officials. During the interview sections of the workshop, the residents would emphatically suggest to the students to put *everything that was possible* in the plan and in accordance with the municipal laws. For example, as one student noted that it was possible under the current CRZ laws to have biogas plant in the fishing village or near the mangroves, one of the residents responded by saying, "Then just put it there! Show it there! Then later we can see whether we need it. Just put whatever is possible for now."

Far from a flawed participatory exercise, these encounters between the students and the fisher community show the different readings and potentials that are ascribed to a plan, the many ways in which urban planning discourses and planning practices intersect with the public reading of these practices. They demonstrate the ever-present possibility for things to not go as planned (Mosse 2003). For the residents of Malvani village, the purpose of this participatory exercise was *the creation of a plan as an image*. The point of the plan was to maximize their potential to claim rights and amenities under the current CRZ and municipal laws. The plan was important to the fisher community's political activism and position.

For KRVIA and TISS, the plan was not just an image, but also a readable document that built a critique of top-down planning methods. This difference was most apparent at the time when the villages received a copy that compiled the studio exercise. It contained the base surveys, drawings that detailed existing land use, land-use change, historical development, street sections, and a conceptual plan that put forward some possible proposals for developing the village. The conceptual plan was put at the very end of the document, which suggested that the process of intervening in the urban fabric required close engagement and a nuanced understanding of a community's needs. During one office meeting in Malvani Koliwada, a member of the cooperative was thumbing through the document. He asked one of the other members to help him find the plan. When they found it at the very end, they immediately suggested separating it from the rest of the document to make copies of the plan in order to show it to municipal corporators. As one of the members looked at the plan, he said, "Once the corporator sees the plan, he will realize that we also know what is possible."

The colleges focused on the information perceived from the plan-as-document while Kolis focused on plan-as-image. While the colleges expected the viewers to construct history, understand spatial complexity, and get a sense of the context before presenting interventions, the Kolis were interested in conveying that they had a plan, they could speak back to the state or developers in the official language, and that since they had the plan, they could not be displaced. Their differing view is better understood as a response to the urgency created by the threat of displacement and, akin to what George Jose (2015) notes about housing conundrums in the Vasai-Virar region, outcomes of changing policy terrains and real estate pressures that generate distinct urban imaginaries. The plan-as-image would immediately place the community in an advantageous position in their struggle for housing. Moreover, these different readings and values should not be taken to mean that the Kolis wanted to ignore historical and social complexities. Neither should it be taken to mean that the colleges were not aware of the land pressures the fisher community faced. Indeed, each was acutely aware of the other's relationship with the plan, and ironically, both these ways of looking at a plan come out of the same historical context and political shifts. Their intersection is most apparent in the public movement demanding participation in the production of the city's new DP.

These different readings of the plan become clearer when contextualized in relation to recent events and aspects of Mumbai's spatial politics. First, the Winter Workshop and the alliance between the colleges and the fisher community must be understood in relation to the participatory planning movement that emerged with the DP's revision. This project was one of many community-led planning projects happening at that time. The critique of top-down planning and mode of conducting studio exercises also came out of the shift toward decentralized planning practices that took place in the late 1990s. Besides making space for partnerships between real estate agents and community members, the move toward decentralized planning practices made space for a new kind of academic practice in Mumbai that was based on interactions between residents and architecture schools or research institutes. During the last few decades organizations such as the Partners for Urban Knowledge, Action and Research (PUKAR), UDRI, and institutions like KRVIA and TISS began undertaking workshops and publishing research projects that proposed a model of working in the city, and of thinking about the urban landscape as a site of work (Paul et al. 2005). Thus, both institutions were keenly invested in a slow and sensitive engagement with the milieu.

To further understand why community-led plans, which lack official capacity, have such political value, it is particularly important to trace its relation to the slum rehabilitation policy. These pedagogical moves and alliances owe much to the grassroots collaborations that were forged as a response to slum displacement and rehabilitation programs. With decentralized urban governance, local NGOs could take an active role in securing housing rights for residents of informal settlements. In the context of this struggle for housing, the creation of cartographic plans and architectural design were mobilized as important technical instruments that gave residents a greater control over the rehabilitation process, helped maintain the community fabric, and ensured access to infrastructure. Participatory exercises and community plans such as the one created in Malvani Koliwada are linked to these rich histories of plans and planning in the city. It is also important to remember that the fishers' need for creating plans, was, in part, to reduce the risk of being labeled slums and moved to rehabilitation housing.

ONE OF THE MOST
IMPORTANT CAMPAIGNS IN
THE RUN UP TO THE
REVISION OF THE DP WAS
THE OPEN MUMBAI PROJECT,
LAUNCHED IN MARCH 2012 BY
THE OFFICE OF P. K. DAS, A
PROMINENT ARCHITECT. THE
PROJECT WAS EXHIBITED IN
THE PRESTIGIOUS NATIONAL
GALLERY OF MODERN ART.

ONE AD FEATURED A PICTURE
OF A SMOG-COVERED
SUBURB. ANOTHER AD, THAT
FEATURED A PATCH OF BRIGHT
GREEN GRASS, BEMOANED
THE LOSS OF THE CITY'S
GARDENS TO INFORMAL
SETTLEMENTS, OR
"DRUGGIES AND LOVERS,"
AND COVERED IN WASTE.

IN THE WORDS PLASTERED
ON THE GALLERY WALL, THE
PROJECT WAS AN ATTEMPT
TO RETHINK THE FABRIC OF
THE CITY, TO TRY AND PLAN
IT BASED ON A "SYSTEM OF
PUBLIC SPACES WITH EQUAL
ACCESS TO ALL," AS
PLANNING, URBAN DESIGN,
AND ARCHITECTURE WERE
"TOOLS FOR INFLUENCING
SOCIAL CHANGE."

I WAS AT THE EXHIBITION WITH DEV KOLI, THE HEAD OF THE FISHING COMMUNITY AT MALVANI KOLIWADA. HE HAD CALLED ME THE NIGHT BEFORE AND SAID THAT MANY FISHERS WOULD BE GOING TO THE EXHIBITION. WORD WAS THAT IN THE MAPS DISPLAYED IN THE EXHIBITION, SEVERAL FISHING VILLAGES WERE DEMARCATED AS SLUMS.

ACCORDING TO DEV KOLI, THE FISHING COMMUNITY WAS ANXIOUS AS P.K. DAS'S PROJECT ENVISIONED A MASSIVE REDEVELOPMENT OF THE CITY'S COAST AND SEVERAL PROJECTS WITHIN IT WERE PROPOSED ALONG STRETCHES USED BY FISHERS.

ON THE FIRST FLOOR OF THE EXHIBITION WAS A SET OF SATELLITE MAPS THAT SHOWED THE EXISTING CONDITION OF THE CITY. THE AERIAL VIEWS WERE IN MUTED SHADES OF BLUE, GREEN, GREY, AND BROWN AND OVER THIS THE AUTHORS HAD JUXTAPOSED BROAD CATEGORIES OF HOW THE CITY'S LAND WAS BEING USED.

"IS THIS THROUGH THE GOVERNMENT OR WHAT?" MR. RUPESH, A REPRESENTATIVE FROM ANOTHER VILLAGE ASKED.

IN HIS HAND HE HELD A CAMERA THAT WAS POINTED AT THE MAP. HIS EYES WERE ON THE VIEWFINDER, WHICH WAS ZOOMED INTO A DETAIL. I HEARD THE CAMERA CLICK.

"THE CHIEF MINISTER HAS INAUGURATED IT SO EVEN IF THE GOVERNMENT HAS NOT STARTED IT, IT [THE STATE] WILL DO SOMETHING WITH IT . . . HOW DID THEY GET THIS DATA? HOW DID THEY MAKE THESE MARKS?" HE ASKED, OBVIOUSLY REFERRING TO THE AREAS THAT WERE MARKED AS SLUMS.

IT DOESN'T LOOK LIKE AN IDEA. EVERYONE HAS IDEAS . . . BUT TO PUT IT ON A MAP AND SHOW HOW TO DO IT, THAT IS ANOTHER THING.

NOW THERE IS THIS EXHIBITION. EVERYONE HAS SEEN IT. IT IS IN THE NEWSPAPER. SO IF SOME DEVELOPER OR CORPORATOR SEES IT, ANYTHING COULD HAPPEN.

THEY MIGHT JUST USE IT...THE MAPS AND PLANS ARE ALREADY THERE... AND IF THE DEMARCATED AREA HAS NOT BEEN SHOWN PROPERLY, THEN WHO IS GOING TO STOP AND CHECK?

ANOTHER PART OF THE EXHIBITION REIMAGINED PUBLIC SPACES ALONG THE CITY'S COAST. EACH PROJECT HAD PHOTOGRAPHS DESCRIBING THE CURRENT CONDITION OF THE URBAN EDGE. RIGHT NEXT TO IT WERE EXAMPLES OF PUBLIC SPACES FROM ACROSS THE WORLD, AND IMAGES OF HOW THAT SPACE COULD BE CHANGED BASED ON THE INTERNATIONAL EXAMPLES. WE EXITED THE BUILDING AND WALKED OVER TO A TEA STALL ON THE ROAD. A GROUP BEGAN A VOCIFEROUS DEBATE ABOUT THE EXHIBITION AND ITS IMPACT ON THE FISHER COMMUNITY. MR. VAITY, MEMBER OF THE FISHER COOPERATIVE AT MALVANI, WAS WONDERING ALOUD ABOUT THE STATUS OF UNOFFICIAL OR CONCEPTUAL PLANS.

THEY WILL TAKE OVER THE LAND IN THE NAME OF BEAUTIFICATION . . . I DON'T THINK THAT THIS IS JUST AN IDEA.

ONE CANNOT PREDICT WHEN SOMETHING GOES FROM AN IDEA TO AN ACTUAL PROJECT.

IF SOMEONE SEES A BEAUTIFUL BRIDGE IN THE MANGROVES WHEN THEY GO ON THEIR INTERNATIONAL HOLIDAY, THEY WILL THINK, LET'S BUILD A BRIDGE IN THE MANGROVES IN MUMBAI! IT CAN HAPPEN ANY TIME.

The Lives and Times of Plans

In 1995, the state government of Maharashtra, India, deregulated its slum reha-bilitation policy. It established the Slum Rehabilitation Authority (SRA), which would provide the legal framework for rehabilitation, but the process itself would be undertaken through partnerships between residents and private develop-ers who would receive development incentive. This landmark reform radically altered the lives of millions of residents living in the city's informal settlements. The state promoted the enterprise as one that would give slum dwellers greater control over the construction, design, and location of the rehabilitation schemes to which they were displaced. However, the years following the institution of the SRA saw a sharp rise in demolition drives—over ninety thousand houses were demolished in 2004–5 (Dupont et al. 2013). The SRA became the most direct route for residents of informal settlements to gain housing, security, and recog-nition as rights-bearing residents of the city, via "voluntary" participation in a violent relocation process (Bhide 2023).

In formulating the SRA, the state authorities of Maharashtra cast all forms of informal housing as illegal, unsanitary, and environmentally harmful zones that warranted active urban interventions. By moving slum dwellers to apartment-based schemes, state interventions claimed to provide them with better living conditions, access to domestic infrastructure, and legalized housing while ensur-ing better environmental conditions in the city and protecting the rights of its taxpaying residents. In a detailed account of the history and working of the SRA, Vinit Mukhija (2000) shows how, over the years, the slum rehabilitation process was restructured as a public-private partnership. This resulted in new institu-tional relations between slum communities and other parastatal agencies, who at first stepped in as advocates for residents' rights in the rehabilitation process.

Apart from the new institutional frameworks, the SRA also made room for new participatory design practices to emerge.[4] For instance, the Society for the Promotion of Area Resource Centers (SPARC), an NGO working on slum dwell-ers' right to housing, grew to prominence in its role as a mediator between slum cooperatives and private developers. SPARC was not only involved in helping communities secure housing, but also organized avenues for community-led design. These alliances gave residents a means to engage state officials and tech-nical experts, and in the process, planning, and the state-society interactions it fostered, emerged as a means to negotiate political claims and to affirm citizen-ship (Patel et al. 2002).

The SRA is a double-edged sword when considered in relation to the koli-wadas and the fisher community's struggle. The 2011 CRZ's housing reforms, which raised the possibility that koliwadas could be lumped with other informal settlements and labeled as "slums," was a source of great anxiety given the history, structure, and process of slum rehabilitation in Mumbai. However, the SRA also provided the language and structure for creating alliances and forging a dialogue through design. The DP's revision complicated the question of the Koli community's land rights and raised the possibility of displacement. The studio project at the center of the alliance between the koliwadas and the educational institution was timely because it provided the Koli community the possibility of interfacing with state officials to intervene in the DP's revision, despite the fact that the plans that the students produced were entirely unofficial. Similar to P. K. Das's "Open Mumbai" project, it could move from the realm of ideas or concept to concrete action, given that it was a technical drawing.

Technical images are key to the alliances and networks that sustain participatory planning. These instruments act as a common ground—communities work with plans, models, and maps as a means of reimagining their neighborhoods and use those drawings to negotiate with the state and with developers. Participatory initiatives are not always about countering the state or resisting development: communities such as the fishers are clearly invested in working from within the technical language of the state's cartographic drawings. For the fishing communities, the very *creation* of the plan (especially one that follow's the official visual convention) imparts the potential to make claims to infrastructure, housing rights, and amenities. It also becomes a means of protecting themselves from private developers, despite the knowledge that the MCGM may not accept a conceptual plan. It becomes a means of establishing (or at least showing the possibility of establishing) a consensus within each village and between fishing villages located in different parts of the city.

Thinking about the participatory exercise and the kinds of urgency these drawings compel or respond to, I am drawn to Namita Dharia's (2022) work on the relation between the ephemeral and the permanent in the architectural industry. Like the drawings at a construction site, development plans too are artifacts that are instrumentalized in the service of creating an enduring city. And yet, they are not one fixed, enduring thing—the plans take various avatars and have different statuses before they can manifest in material terms. These "unofficial" plans are drawn not just in relation to those possible manifestations, but also in relation to the various life stages of official plans—they are a means by which koliwada residents can "time" the state to capture political and developmental possibilities.

An official plan goes through several stages before it comes into force. During the revision process, the plan exists as a draft. Once a draft is completed, it is released for suggestions and objections from the public, after which it is revised again. Once it is passed, it has legal status for a specified time, and unless it is extended, it expires. Even in the case of planning institutions, the line between official and unofficial is not a fixed one, and this is one reason why a conceptual plan has political capacity and can be used to make land claims. For instance, the first master plan document for Mumbai, *An Outline Master Plan for Greater Mumbai*, was prepared by Albert Mayer and N. V. Modak, and released in 1948. The Modak-Mayer plan, as it came to be known, had no official status and neither did the Municipal Corporation adopt it in any official capacity. Yet, its suggestions show up in several municipal projects and in many subsequent plans (Dossal 2005; Dwivedi and Mehrotra 1995). The first official plan for the city was proposed in 1964 (and released after a three-year delay), nearly two decades after the Modak-Mayer plan was released. The *Development Plan for Greater Bombay* restructured the city by zoning new residential, industrial, and commercial areas, and by dispersing the city's growing population northward, to the suburbs (Shaw 2004). The second Development Plan was prepared in 1981 but was delayed for more than a decade, as the Municipal Corporation only released it in 1994. Apart from these delays, several planners have commented on the "partial" or "faulty" implementation of the city's plans (Phatak and Patel 2005; Nallathiga 2009). This lineage of plans and planning documents undoes ideas that take the official and the unofficial, or formal and informal, as mutually opposed, exposing their slipperiness. It positions informality as an inherent feature of urban planning and policy (Gandolfo 2013; Roy 2005). The relationship between the formal and informal is particularly interesting in the case of the DP's revision process. When the revised DP was finally released in 2014, it was met with widespread opposition, and under public pressure, the MCGM scrapped the new DP and restarted the revision process.

Another reason why community planning gained political traction was due to the introduction of e-governance initiatives at the municipal and ward level, which were intended to restructure the relation between different stakeholders such as the citizens, politicians, and state officials. Ideas like increasing accountability, striving for transparency, reducing corruption, resolving poverty and inequality, and enabling the free flow of information between government and citizens underlie all these initiatives. As Mazzarella (2010, 783) writes, these technological initiatives were promoted as the means of solving pressing issues like urban poverty by "bridging the 'digital divide' between the wired and the unwired, between global cities and information slums." In Mumbai, these initiatives coincided with several infrastructure projects that displaced thousands of people, and events like the 2005 Maharashtra floods that put additional pressure on communities living in informal housing. Consequently, public participation, access to technology, and open governance were upheld as the way forward. In Mumbai, after the 2005 floods, which were seen as the direct outcome of a flawed planning, urban activists suggested a critical revision of the DP as a means of addressing the city's infrastructural woes and the problems of those who were at risk—people in informal settlements (Bhagat et al. 2006). While equity in planning, transparency, and the free flow of information were guiding principles for the community initiatives that happened around the DP's revision, they were not without their inherent tensions and problems. For example, the Open Mumbai project pitted groups like the fishermen in opposition to other residents who had different visions for the city's shoreline as a space for recreation rather than as a space for work. The Open Mumbai project captured the public imagination not just because of the information contained in the drawings, but because it followed the language of official plans.

Writing about cadastral surveys in Paraguay, Kregg Hetherington (2011) notes that development experts use the idea of access to information as a means of exerting control. Hetherington argues that while the idea of free moving information is described as having great promise in development circles, it overlooks the material forms of the documents and artifacts that contain the information. Anthropologists of paper and the everyday life of bureaucracies (Hetherington 2011; Mathur 2016) distinguish between information and the material artifacts on which data appear, and pay close attention to the values and forces attributed to documents because of the marks, signatures, stamps on it, or their material conditions, or their availability. Second, while programs like the CRZ evoke ideals of transparency, they do, as Nayanika Mathur (2016) writes, rely on paperwork to produce that bureaucratic transparency. As a result, documents end up gaining value, and in a situation where the policy is nascent, a heightened sense of anxiety surrounds them.

As I found out later, Dev Koli and Rupesh had an extensive archive of documents that they were constantly adding to. These archives were very loosely held—variously tucked away somewhere as set of sheets in the fishing cooperative's office, or even a file or a set of loose sheets kept inside a cupboard, or as random files stored in a computer with no apparent order. These archives were built with the sense that any document had the potential of becoming important in the event of a change in the policy or situation. When collecting copies of documents that could be used for official purposes, Dev Koli would ensure it was at the right resolution and had the relevant signatures and details. Though entirely unofficial, these archives and the desire to accumulate documents are outcomes of bureaucratic histories—of exerting and gaining control and power through forms of writing and paperwork, or even obfuscating writing and denying paperwork (Gordillo 2006; Raman 2012). They are connected to the fraught histories of documenting and governing the poor, and the provisional ways in which statuses and identities are granted to people, and in this case, to places as well (Sriraman 2018).

In these archives, plans had special value. Both Dev Koli and Rupesh collected all kinds of plans, including ones that are publicly available, or ones that had no relation to their case. To recall what Mr. Vaity said outside the gallery, plans carry a concrete possibility of being filed, registered, and eventually being undertaken as an enterprise. Moreover, plans are documents that are notoriously difficult to obtain from government archives. As a result, they come to occupy a special place within this archive and are of great significance to claiming rights. Though important, these plans were not the only objects vital to land politics. Given the importance of proving identity into claim land rights, plans were used in tandem with a range of pictures and objects that confirmed the community's indigenous status. This was most apparent at the seafood festivals where the community reestablished its identity by drawing on the visual history of depicting Kolis, particularly Koli women, as native subjects.

MUMBAI'S BHAU DAJI LAD CITY MUSEUM HAS A LARGE COLLECTION OF CLAY FIGURES AND BUSTS FROM THE LATE NINETEENTH AND TWENTIETH CENTURIES. THIS COLLECTION IS STORED ON THE UPPER LEVEL OF THE BEAUTIFUL MUSEUM, AND THE FIRST MODEL THAT GREETS VISITORS IS A CLAY BUST OF A KOLI FISHERMAN.

THE ADJOINING TEXT, ADAPTED FROM *PEOPLES OF BOMBAY* (STRIP AND STRIP 1944, 20), DESCRIBES THE KOLIS AS "ONE OF THE FEW ABORIGINAL TRIBES OF *DRAVIDIAN ORIGIN* . . . AND HAVE CHANGED LITTLE SINCE THEY ARRIVED IN HEPTANESIA. . . .YOU CAN SEE THEIR SETTLEMENTS IN COLABA, WORLI, MAZGAON AND THE NEARBY SUBURBS OF BOMBAY."

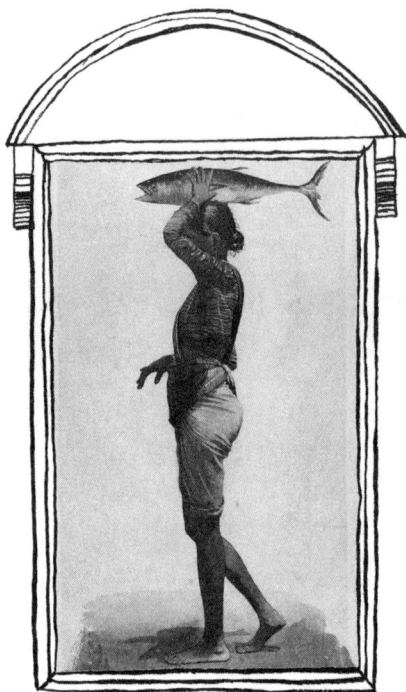

PARTHA MITTER (1994) WRITES THAT IN THE EARLY NINETEENTH CENTURY, AFTER THE ESTABLISHMENT OF THE EAST INDIA COMPANY IN BENGAL, LOCAL ARTISTS WERE COMMISSIONED TO PRODUCE WATERCOLORS, DRAWINGS, AND CLAY FIGURES IN WHAT CAME TO BE CALLED THE COMPANY STYLE.

ALONG WITH TOPOGRAPHICAL, ARCHITECTURAL, AND ARCHEOLOGICAL DRAWINGS, ARTISTS ALSO PRODUCED ETHNOGRAPHIC SETS, WHICH USED VISUAL SIGNS LIKE CLOTHES AND TOOLS TO MARK CASTE IDENTITY ON THE SUBJECTS.

IN 1865, AN ARTIST NAMED JOHN GRIFFITHS ARRIVED FROM SOUTH KENSINGTON TO TAKE CHARGE OF THE PAINTING DEPARTMENT AT THE J.J. SCHOOL OF ARTS IN BOMBAY. IN 1872, GRIFFITHS PAINTED A WATERCOLOR SKETCH TITLED *WOMAN HOLDING A FISH*.

AMONG GRIFFITH'S MOST FAMOUS STUDENTS WAS MAHADEV VISHWANATH DHURANDHAR, WHO LATER BECAME THE FIRST INDIAN DIRECTOR OF THE J.J. SCHOOL. DHURANDHAR PRODUCED SEVERAL DIFFERENT PAINTINGS OF KOLI WOMEN THAT WERE INSPIRED BY GRIFFITHS'S OWN, ALONG WITH PORTRAITS OF PEOPLE BELONGING TO OTHER COMMUNITIES FOR ILLUSTRATED ACCOUNTS OF THE PORT CITY.

DHURANDHAR'S WORK WAS PIVOTAL TO THE CREATION OF A REGIONAL IDENTITY (JAIN 2007; MITTER 1994).

THE KOLI WOMAN HAS EMERGED AS AN UNMISTAKABLE SIGN FOR MUMBAI'S "CHAOTIC AND MONGREL WORLD" (PRAKASH 2010, 6).

A HISTORIC CHAIN LINKS GRIFFITH AND DHURANDHAR TO POPULAR DEPICTIONS OF KOLI WOMEN IN MARIO MIRANDA'S CARTOONS. THE CHAIN EXTENDS TO DEPICTIONS OF KOLI WOMEN AS OBJECTS OF FANTASY IN BOLLYWOOD FILMS, THE MOST FAMOUS BEING MADHU DIXIT'S DANCE NUMBER IN THE MOVIE, SAILAAB.

IN 2011, WHEN MUMBAI HOSTED ITS FIRST COMIC CON, ONE OF ITS OFFICIAL MASCOTS WAS WONDER BAI, A VERSION OF WONDER WOMAN AS A KOLI FISHER.

Contours of a Dispute

Well before the road turned toward the fishing settlement, a large sign announced the Versova Seafood Festival. The sign had a large photograph of Raj Thackeray, the leader of the Mahararashtra Navnirman Sena (MNS), a right-wing, regional political party. The reflection on his sunglasses caught the crowd of people who often come to witness his fiery speeches. Under Thackeray's image were rows of headshots of party workers and local political representatives and the text on the poster welcomed all seafood lovers to the festival. I found similar posters as we were walking toward the festival; the same image made an appearance at regular intervals, tied to lampposts and gates. At the end of JP Road, near the site of the festival, there was a similar poster printed by the Shiva Sena. On this poster, photos of the late Shiva Sena supremo Bal Thackeray and his son Uddhav Thackeray (the current chief minister of Maharashtra) urged their viewers, *Radaicha nahi, ladaicha!* (We must not cry, [we must] fight!).

In 2008, MNS chief Raj Thackeray inaugurated the Koli Festival at Dharavi Koliwada. Raj Thackeray established the MNS in 2006, following a split with Shiva Sena (which was run by his uncle, Bal Thackeray, until his death in 2012). Both parties champion the rights of the *Marathi manoos* (Marathi people) and oppose migrant workers. Raj Thackeray's inaugural speech at Dharavi Koliwada was a familiar reflection of his party's xenophobic ideology. The speech urged the Kolis to take an active part in their community's development:

> If anyone talks about rights over Mumbai city, then the ones who have the most valid claim are the Kolis. . . . The other day a number of Koli mothers and sisters came to my house and complained that the "bhai-yas" had taken over their business and they did this and that.[5] . . . But I ask, 'How could they take it?' It is because you have made a lot of money buying and selling fish over the years and have become rich. Now you are no longer willing to sell fish, you are ashamed. Consequently, someone else has come to take your work. But I say—do honest work and then come to me, I will support you and provide you with whatever you need.[6]

When the MNS organized the *Mi Marathi Khadyamahotsav* ("I am Marathi: Food Festival") in 2010, it was widely regarded as a way of broadening their appeal among the city's middle-class Maharashtrian population (Packel 2011). In the case of the seafood festivals, political support seemed to run both ways as the festivals became important sites where the Koli community sought support for its own cause—especially after the 2011 CRZ notification. These well-attended events became a means of garnering both political and popular support for the koliwadas that organized them.

One of the fallouts of the 2011 CRZ notification was that it deepened the schism between the "local" Koli population and those were termed as "outsiders." D. Parthasarathy (2011) explains this conundrum in his article on urban development and the interests and rights of the "hunters, gatherers, and foragers" in Mumbai. The interests of communities such as Kolis who live by catching fish are often not included in Mumbai's urban development plans. However, the process of claiming rights for native communities is often based on a direct opposition of migrant communities. These migrant others include people who have been staying in the community for several years and are, for all other purposes, considered a part of that community. This ethnic mobilization revealed itself several times in the short speeches that were given during the Versova Seafood Festival, which exhorted absent government officials to remember the city's history and recognize their claim. Simultaneously, the claim was manifest in images, signs, advertisements, and decorative objects at the festival.

The space of the Koli festival shows how indigenous identity—and the act of claiming it—is not based on one fixed thing. Instead it can be thought of as a coming together of many moving parts. The digitally created images along with objects like fishing nets and replicas of fish, placed the Koli cause within the broader framework of regional politics in Maharashtra. At the seafood festival, the visitors' cone of vision was filled with pictures of Koli men and women, nets, and fish, along with large portraits of political figures like Bal Thackeray and Raj Thackeray. As these different elements came together in the viewer's gaze, they articulated the Koli community's indigeneity and their land claims. At the same time, when combined with images of political patronage, the visual montage joined the Koli struggle with regional politics. The signs with the festival sponsors, which included the Oil and Natural Gas Corporation of India, added economic support to their claims. As much as these visuals pulled an abiding identity and history into the present, they also projected it into the future: by gathering public sentiment and political support, and by reminding the city of its history, it set the stage for the Koli community to seize the developmental opportunities of the CRZ and claim their right to have a say in the city's transformation. It let them claim village land while separating themselves from other residents of informal settlements.

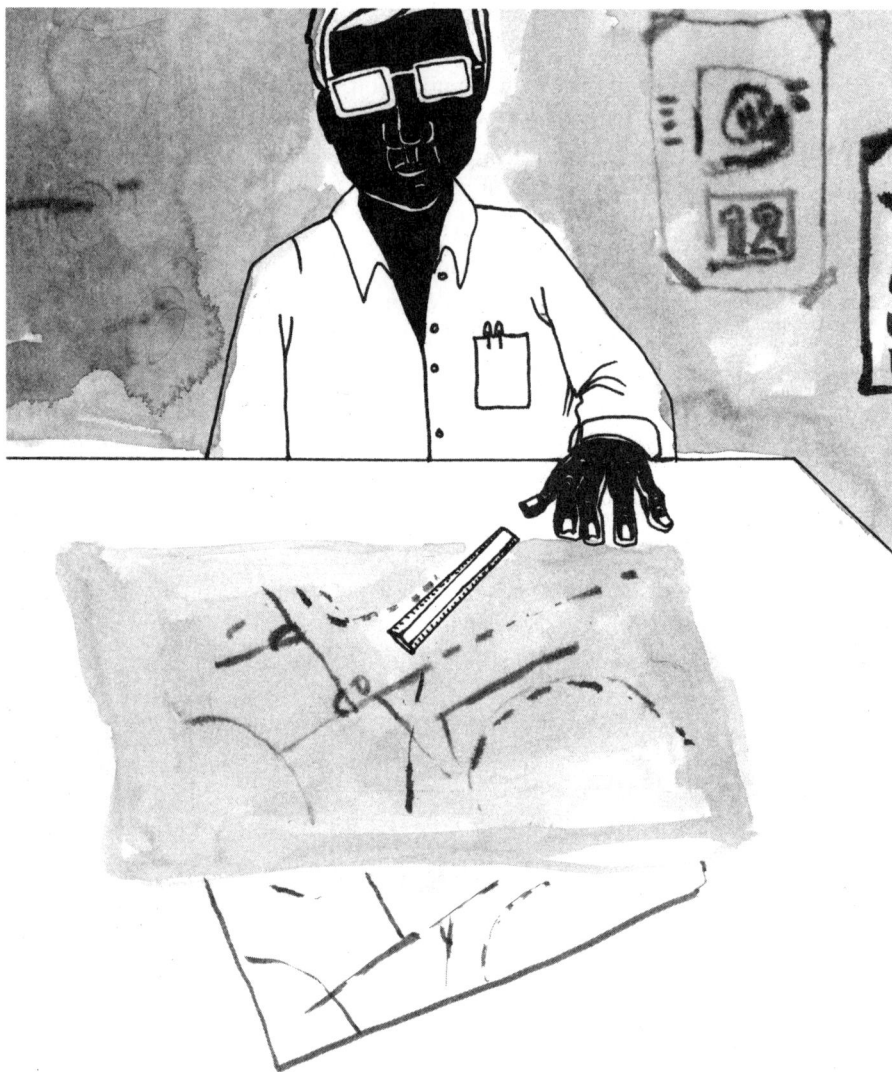

IN THE WINTER OF 2011, ONE OF THE WHOLESALE DEALERS IN THE FISHING VILLAGE TOLD ME THAT HE WAS FEUDING WITH HIS NEIGHBOR OVER A PROPERTY THAT HIS FAMILY OWNED.

HE GAVE ME VAGUE DETAILS: HIS FAMILY HAD TASKED HIM WITH DEVELOPING THE PROPERTY IN EXCHANGE FOR AN UNMENTIONED SHARE OF THE PROFITS, AND HE HAD HIRED A DEVELOPER. UPON VISITING THE SITE, THE DEVELOPER CLAIMED TO HAVE FOUND THAT THE NEIGHBOR HAD CONSTRUCTED A STAIRCASE, WHICH INFRINGED ON PROPERTY. AFTER FAILED INFORMAL NEGOTIATIONS, HE DECIDED TO CONSULT THE SERVICES OF A GOVERNMENT SURVEY OFFICER TO RESOLVE THE PROBLEM. CURIOUS ABOUT WHAT THE SURVEYOR WOULD DO, I ASKED TO TAG ALONG.

WE MET IN THE DEVELOPER'S OFFICE WHERE THE SURVEYOR AND HIS ASSISTANT WERE GIVEN FOOD AND DRINKS. AFTER THE MEAL, THE SURVEYOR PRODUCED A SHEET ON WHICH HE TRACED A PART OF THE CADASTRAL SURVEY, AFTER WHICH THEY WERE LED TO THE PLOT WHERE AN OLD, DILAPIDATED HOUSE STOOD.

ONCE AT THE PROPERTY IN QUESTION, THE ASSISTANT SET UP THE PLANE TABLE ON WHICH HE
AFFIXED THE TRACED SHEET. OVER THIS, HE LAID ANOTHER SHEET AND BEGAN THE PROCESS OF
RESURVEYING THE PROPERTY ON THE GROUND.

THIS PROCESS WAS LIKE A SURVEY IN "REVERSE" AS IT TOOK THE EXISTING SURVEY RECORD AND
TRACED IT BACK ON THE GROUND TO CHECK THE EXTENT OF PROPERTY. THE SURVEYORS FIRST
ESTABLISHED THE BOUNDARY OF THE PROPERTY BY CONSULTING THE DRAWING AND MARKED IT
ON THE GROUND WITH POLES AND FLAGS. SIMULTANEOUSLY, THE ASSISTANT CREATED ANOTHER
DRAWING ON THE PLANE TABLE, WHICH SHOWED THE POSITION OF THE NEW CONSTRUCTION IN
RELATION TO THE OLD PLOT LINES. AFTER FINISHING THE DRAWING, THE ASSISTANT CAREFULLY
OVERLAID A TRACING OF THE OLDER SURVEY TO UNDERSTAND WHETHER THE PROPERTY
EXCEEDED THE PLOT LIMITS.

I SAW THE SAME SURVEYOR RESOLVE A DIFFERENT DISPUTE BETWEEN TWO VILLAGERS WHOSE FIELDS ABUTTED EACH OTHER, WHERE ONE PROPERTY OWNER ALLEGED THAT THE OTHER HAD EXTENDED THE BOUNDARY OF HIS PROPERTY INTO HIS FIELD.

EVERY TIME, THE SURVEYOR WOULD RE-CREATE THE SURVEY ON THE GROUND AND VERIFY IT AGAINST THE OLD RECORD TO CONFIRM OR DISPEL THEIR SUSPICIONS. THE PROPERTY OWNERS WOULD THEN RESOLVE THE MATTER BETWEEN THEMSELVES.

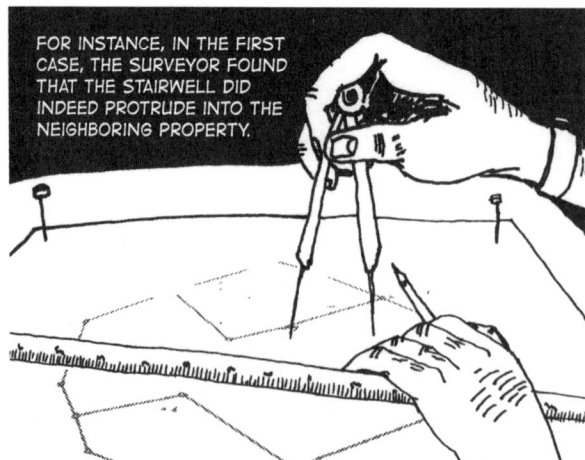

FOR INSTANCE, IN THE FIRST CASE, THE SURVEYOR FOUND THAT THE STAIRWELL DID INDEED PROTRUDE INTO THE NEIGHBORING PROPERTY.

BASED ON THE DRAWING, THE TWO OWNERS NEGOTIATED A DEAL WHERE THEY AGREED TO SHARE THE STAIRCASE.

A Historical Record that Charts the Future

At the Versova Koliwada Festival, there was a poster that faced the entrance of the fishing village. The poster, printed by the Association of Women Fish-Sellers of Marol Bazaar, had a blue background of a beach scene with words written in bright yellow. As with the speeches in the festival, the words were addressed to the nameless government surveyor who would come to determine the status of the village:

> *Naka deu dhamkya—naka karu prant vad!*
> *Naka karu rajkaran!*
> *Mumbaicha 7/12 kora ka? Kolyanche nav dya!*
> *Mumbai ahe Kolyanchi, nahi konache bapachi!*
>
> Do not give threats—do not resort to communalism!
> Do not do politicking!
> Is Mumbai's 7/12 blank? Give the Koli peoples' names!
> Mumbai belongs to Kolis, and not to anyone else!"

At the bottom left corner of the poster was an inset of a map with a red dot signifying the location of the fishing village. As I stood reading the poster, a young man standing near a small truck called out laughing, "Madam, the koliwada is over there!" I made to leave when an older man who was also watching me said, "Go inside and you will see that this is our village. This is a place where Koli people live. This is not a slum."

The *saat-bara* extract (also referred to as saat-bara, or simply 7/12), which was frequently mentioned in several speeches during the festival and at different points in my field work, is a type of property record that is maintained by the Revenue Department of the Maharashtra state. It provides details of agricultural land or land that was converted from the agricultural to the nonagricultural category. Although the saat-bara is no longer used as a property record (for nonagricultural land, at least) in Mumbai, the term is used in common parlance in Marathi when referring to the property card, or simply to mean, "survey." In this context, the term is used in relation to the survey the MCGM intended to conduct to identify the existing koliwadas in the city. When a poster or a representative demands that government officials "show the koliwadas in the *saat-bara*," they mean that all the fishing settlements must be marked and recognized as koliwadas, Kolis must be given access to land for housing and for fishing, and most important, that these spaces (and by extension the communities) must not be labeled otherwise—as slums.

The actual saat-bara record has many parts and details. Apart from specifying ownership, it provides information such as the kinds of crops sown on agricultural land, quantity of produce, and details of taxes paid, and court cases related to that specific plot to which it refers. Attached to each saat-bara is also the *ferfar* (changes) report that gives details about the history of the land and changes in ownership. In many ways, the Koli's claim for land rights is akin to looking up the saat-bara of the land in question and opening the ferfar report where the Kolis emerge as the first inhabitants, thus establishing their claim over the settlement, resources, and development opportunities. In this sense it revives the historical narrative of the city in which the Koli's were the first to inhabit the islands on which the metropolis of Mumbai grew. Evoking the saat-bara, in combination with signs that identify Kolis as natives, is an important means of intervening in the moment of the survey, and of addressing a mythical state surveyor to control how names, identities, and categories get recorded and appear on the final survey sheet.

In a manual on city surveys written for government officers, author A. K. Gupte (2009) remarks that pieces of land used for agricultural purposes are invariably sold off and used for residential or commercial purposes. The moment this happens, Gupte writes, "Then the City Survey law is, as it were, lying in wait for them, and the individual plots so formed in the blanks or islands left in the original maps (City Survey Sheets) will at once be mapped in detail." Once a plot of land is used for anything other than agricultural purposes, it passes into the City Survey register, entering the realm of the urban. However, this work of transforming agricultural land for urban use is not done by the surveyors, landowners, and developers alone, but also through the actions of the survey itself: Gupte's manual evokes a picture of the survey as an agent that lies in wait and pounces when the opportunity presents itself to turn everything into the city.

The survey derives its agency not just from what it depicts but also its local lives and histories. Its role as a mediating agent shows how the moment or process of drawing a survey can become a site for negotiation. Via its relationship with other images, it becomes a powerful agent for articulating indigeneity and forging political alliances. Surveys and plans also share a complex relationship with time. Different stages of a plan's life offer different capacities for political action. Following the ways in which the Koli community works with plans and surveys offers windows into the temporalities of environmental politics. Though images allow the community to engage other actors in the planning process, it also delimits the frameworks available for claiming rights and the focus of this engagement. The threats posed by the decadal visions of plans—displacement, loss of land, loss of economic opportunity—obscure longer-term concerns caused by coastal pollution, large-scale infrastructural projects, and the effects of toxic waste at the shore.

"THIS IS NOT A DEAD ZONE"

Pollution, Archives of Marine Life, and
Multispecies Encounters

WHEN I FIRST STARTED HELPING SORT FISH, I WOULD SCRIBBLE DOWN THE PIECES OF GARBAGE I SAW AT THE BEACH.

MY LIST WAS MOSTLY A MONOTONY OF PLASTIC. LOTS OF SMALL AND LARGE BAGS, MOSTLY TORN. THERE WAS THE OCCASIONAL BOTTLE, TUBE, OR FRAGMENT FROM A MYSTERIOUS OBJECT. ONCE, WHEN I PULLED OUT A SHREDDED BAG FROM THE CATCH, THE UNFURLING TENTACLE OF AN OCTOPUS GREETED ME. ANOTHER TIME, I SAW A CRAB SCUTTLING AWAY WITH A BOTTLE CAP.

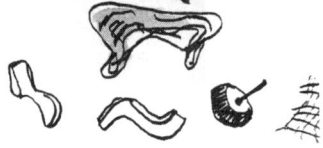

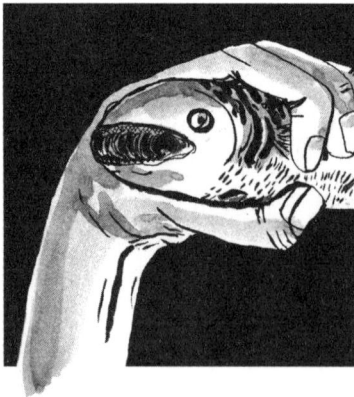

Garbage

Traditionally, the fishing season begins at the end of the monsoon with a prayer requesting the sea to calm its waters and deliver a bountiful catch.[1] The creeks, which meet the sea, offer a safe space to park and launch boats and the smaller one- or two-man vessels use it as fishing grounds as they move in and out of shallow creeklets with ease. They are also channels that convey tons of waste from the city into the sea. The creek, fish, and life and work in the fishing village—and the garbage—are inextricably linked with each other. I worked with the women on the beach sorting and selling fish. Speed was of essence because we had to catch the wholesale dealers at the beach before they finished buying their lots for the day and because we had to beat the rising tide. This work was also how I came to have a very tactile and visceral relationship with garbage.

Malvani is located close to Manori Creek, which forms one of the major estuarine systems of the city. Because of Mumbai's new Development Plan and the coastal policy, the fishing villages became sites of intense real estate speculation. These changes entangled Mumbai's fisher community, the Kolis, in a political contestation, which was not just about rights, but also about the material and ecological future of the city's edge.[2] Given that the fishing villages were located on prime real estate (Warhaft 2001), the public meetings in the villages tended to focus more on the immediate threats of displacement and livelihood loss. However, long-term climate anxieties would erupt from time to time in the everyday work of nearshore fishing. Garbage was the most palpable sign of things to come in the far future.

At present, Mumbai has two working dumping grounds, Deonar and Kanjurmarg, both of which are located on the city's estuarine edge. Deonar, the city's largest dumping ground, is the older of the two and operating well beyond the end of its capacity. The MCGM keeps filing for extensions to keep it open and is also locked in a legal battle over its plans to expand its facilities at Kanjurmarg. The courts ordered a stay on this expansion as it could harm the tidal ecosystem (B. Chatterjee 2019a). The MCGM has long cited the lack of space as a reason to build dumping grounds in the wetlands, which are large areas of marsh and mangrove forests into which the creeks open out and flood with the tide. It promises to minimize damage by using various management technologies while also turning a profit through waste-to-energy programs, both in the new dumping grounds and in closing the older ones, which are to be bio-mined (using microbial action to eat away organic waste and recover metals in the dumping ground).

These sites seek to translate waste economically and materially—from a discarded substance to other things such as energy and land that have much greater value. In the process, waste becomes a part of the energo-political system of carbon credits that are derived from its translation to energy (Boyer 2014). The success of such a translative enterprise depends on controlling how substances mix, and engineering critical separations. However, in the spongy, wet estuaries where these neat separations are fantasized, waste slips, mixes, and merges.

In 2017, I expanded my project to think beyond the coastal policy and more broadly about coastal transformations, climate change, images, and infrastructure. Solid waste management—especially the question of where to put it—was an important part of this research. The new ground—in Kanjurmarg—too was in estuarine marshland, linking it to long histories of dumping and reclamation in Mumbai. In order to think about this tension between mixing and separation, I began talking to waste management scientists. Scientists who work on waste curate the chemical life of dumping grounds by carefully separating and controlling the interactions between different substances. These controlled interactions link waste to global economies of energy and carbon reduction but obfuscate the existing garbage-marsh-people continuities. However, the ways in which scientists engage with the chemistry of substances is revealing of the multiple temporalities in which waste acts. The scientists' understanding of waste opens the possibility of thinking about waste's latencies—the time lag between the moment that waste is dumped and when it eventually reacts with other substances or its surroundings (Murphy 2013).

Around that time, a new marine life advocacy group was conducting shore walks to raise public awareness about the plethora of nonhumans at risk from coastal megaprojects. During these walks, the advocates would point to creatures that persist despite coastal pollution. They would encourage residents to walk along the shore and take pictures of marine creatures and upload it onto a digital database where the creatures are meticulously labeled and sorted. While this archive reproduces modern scientific sensibilities of crafting anthropocentric order onto the coast, the work that these images do is the opposite: they not only provide data about the creature displayed, its habits, and where it was found, but also mark a moment of creating a multispecies relationship that recognizes marine beings as coresidents of the city that exist in spite of, or because of, the chemistry of its shoreline. These walks make visible waste's continuities that reorganize ecologies of a coastal city.

The ways in which the nearshore fishing community, waste management scientists, and marine advocates engage, encounter, and see waste makes room for translating it as not just a commodity, but as a part of the shore. These translations are openings to think beyond large-scale technological solutions and managerial discourses of control and commodification of nature. They also break with nature/culture dichotomies and turn instead, toward the sociomaterial entanglements between matter, beings, and landscape. Waste's entanglements impinge on and alter other entities and beings—including humans (Hird 2013). This profusion of potential confounds the received categories into which infrastructural projects place materials and beings. Far from inert, waste matter acts and reacts, sometimes in imperceptible ways, such as the small, innocuous-looking plastic grains that insinuate themselves into the sediment and masquerade as sand. A geological agent produced by humans, waste has already insinuated itself into the environment where it will continue to remain, far beyond human timelines (Chakrabarty 2009). Its physical presence and interactions, which are often unpredictable, imperceptible, or slow, are creating the coast and its possible futures.

IN THE EARLY 2000S, A TECHNOLOGY PARK CALLED MINDSPACE OPENED IN MALAD, A SUBURB OF MUMBAI. SOON AFTERWARD, THE COMPUTERS AND ELECTRONIC SYSTEMS AT MINDSPACE BEGAN GOING ON THE FRITZ. AFTER SEVERAL MONTHS OF FAILURE, REPAIR, AND REPLACEMENT, AN ENVIRONMENTAL CONSULTANCY FIRM DETERMINED THAT THE DELICATE MACHINERY WAS CORRODING AS IT CAME INTO CONTACT WITH SULFUROUS COMPOUNDS EMANATING FROM THE GROUND (NAIR 2007; SAHU 2007). MINDSPACE WAS BUILT ON WHAT USED TO BE A LARGE MUNICIPAL DUMPING GROUND, IN MALAD CREEK'S ESTUARINE MARSHLAND.

CREATED IN 1968, THE DUMPING GROUND HAD RECEIVED MORE THAN 1,000 METRIC TONS OF GARBAGE FOR MORE THAN THREE DECADES (ATTARWALA 1994).

High tide line

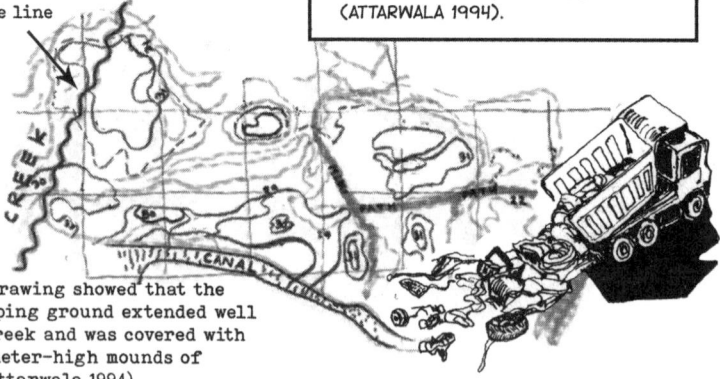

A survey drawing showed that the Malad dumping ground extended well into the creek and was covered with 25-to-30-meter-high mounds of garbage (Attarwala 1994)

THE FILLED-UP MARSH WAS BOUGHT BY THE K. RAHEJA GROUP, WHICH HAD DEVELOPED MINDSPACE AS A BRAND OF COMMERCIAL AND INDUSTRIAL PARKS. MINDSPACE TYPIFIES THE EMERGING ARCHITECTURAL STYLE AND DESIGN VOCABULARY REPLICATED ACROSS INDIAN CITIES (SZE AND GAMBIRAZZIO 2013). ON TOP OF WHAT USED TO BE TOWERING MOUNDS OF GARBAGE, THERE NOW STAND OFFICE COMPLEXES, MANICURED GARDENS, AND ONE OF THE COUNTRY'S LARGEST MALLS.

MALAD MIGHT EVEN BE DESCRIBED AS A SUCCESS WHEN COMPARED TO DEONAR, THE CITY'S LARGEST AND OLDEST DUMPING GROUND. IN 2016, A TOXIC SMOG ENGULFED MANY PARTS OF THE CITY WHEN DEONAR WENT UP IN FLAMES.

THE SMOKE WAS VISIBLE FROM SPACE.

CIVIC OFFICIALS WERE QUICK TO BLAME THE INFORMAL INDUSTRIES THAT OPERATED IN THE GROUND. BUT AS EXPERTS POINTED OUT, THE GROUND HAD BEEN GROSSLY MISMANAGED FOR DECADES AND WAS HANDLING WAY MORE THAN ITS DESIGNED CAPACITY (S. CHATTERJEE 2019).

THE DEONAR FIRE WAS A TURNING POINT IN SOLID WASTE MANAGEMENT PRACTICES IN THE CITY AS THE MUNICIPAL CORPORATION DOUBLED DOWN ON INSTALLING WASTE-TO-ENERGY PLANTS AND ON ITS EFFORTS TO FIGHT ENVIRONMENTAL ACTIVISTS TO EXPAND ITS NEW FACILITIES IN THE KANJURMARG WETLANDS.

12,000

11,000

10,000

9,000

8,000

7,000

6,000

5,000

SOLID WASTE GENERATION IN MUMBAI IN TONS PER DAY (DUTTA AND JINSART, 2020)

99-2000   2004-05   2010-11   2015-16

YEAR

Slow Cooking

Waste's translations—the process through which it appears as or becomes a different thing—are not fixed but occur within a field of possibilities that depends on the timelines, discourses, and frameworks involved. The principles and practices of solid waste management translate substances and spaces classified as waste into commodities of higher value, a process that intensifies entrenched social hierarchies (Gidwani 2015; Mirza 2019). Such a translation depends on keeping matter, beings, and landscape separate and limiting their interactions (Reno 2016). This involves both physical separations—like thick barriers of rubble and plastic that separate garbage from marsh—and conceptual separations based on material classification, such as organic, inorganic, toxic, and inert. Within the dumping ground, many of the components that make up the complex amalgam that is waste are thought to react for a certain time after which they are deemed inert, that is, remaining in the ground as benign, self-contained objects.[3] However, as was the case with the Malad dumping ground, things and landscapes rarely remain within neat separations or act in set time frames: cracks, pores, and the slow work of attrition allow for slippages and unanticipated reactions. The steady degradation of materials like plastic, corroding metal, and the toxic sludge trapped under layers of sediment holds the menacing promise of future action. The marsh too is not a passive container; it reacts with and carries waste beyond the dumping grounds and beyond the horizons of development projects.

Depending on the technology involved, these centralized and mechanized waste regimes introduce new kinds of physical and conceptual separations and categories. For example, incineration technology recategorizes the complex mix of garbage into simplified objects for energy production.[4] The resulting economic and political shifts efface both the communities of waste workers and existing ecologies created by years of dumping waste in marshland.[5] Despite these concerns, the repeating refrains of a city swallowed by garbage creates widespread support from middle-class citizens (Anjaria 2009) for large-scale waste-to-energy projects, which result in spectacular landscapes of high-rise housing, malls, and tech-hubs built on the slippery promise of modernity and urban renewal.[6]

I approached Dr. Urmi Mala in the summer of 2017 to understand waste management practices and technologies applied in dumping grounds.[7] Dr. Mala specializes in solid waste management and is an expert on the physical and chemical processes that occur inside these grounds. She started off by explaining capping technology, which is a popular, modern waste management technique that seals off the trash from the surrounding environment and limits any possible human contact. Typically, before garbage is dumped in an area, the ground underneath is prepared with liners and channels for capturing the toxic leachate that the garbage exudes. This creates an impregnable cell, which does not allow the garbage to leak into the ground below. When the ground reaches the end of its life, another cap is placed on top—layers of gravel, silt, and high-density plastic are put over the piles of garbage. The mounds are steam-rolled and compacted after which it becomes saleable land. As Mala explained, most dumping grounds in India did not begin as engineered sites (also called landfills), which means that there are no liners that separate the garbage from the ground below. Yet, as was the case in Malad, these older dumping grounds are routinely capped off, which only serves to render the garbage invisible. "It gives the illusion of separation," Mala said when we first spoke on the phone, "but there is a lot happening underneath."

A scientifically managed dumping ground, as Dr. Mala described it, is an active site, which must be continuously monitored even if it is engineered to be cut off from its surroundings. The physical separation, monitoring, and maintenance practices are crucial for reproducing modern sensibilities regarding the management of contamination and risk by separating waste from society (Douglas 2002). Depending on the architecture and layout, the ground is filled in parts, where matter is segregated based on its composition to minimize damage to the environment and to maximize gains from composting and recycling. Piles must be carefully engineered so as to not slide; they have to be stratified, and the ground's boundaries have to be scrutinized from time to time to ensure that the garbage stays within. Like a pot that is slowly cooking, Mala said, dumping grounds must be carefully attended to; the conditions under which things meet and mix must be meticulously controlled to derive energy from waste.

One of the key aspects of generating energy from waste is to determine its composition in terms of the broad categories of biodegradable waste, paper, plastic, metals, glass, and inert matter.[8] Mala explained that waste has caloric value (waste management is rife with food metaphors: maceration, digestion, and caloric value) and knowing the broad composition of the garbage allows experts like her to calculate the energy that the dumping ground will likely generate over the course of its life. Given that most of what makes up trash in Mumbai is wet, organic waste, it has a low caloric value for waste-to-energy programs that incinerate waste.[9] Waste's composition in these terms—wet, dry, organic, metal—is also telling of how it degrades. In the existing dumping grounds, this mostly wet garbage lies in large mounds inside which there is no exposure to air (described as anaerobic), and undergoes a host of chemical reactions that are, in theory, predictable (Sahu 2007):

$$[\text{organic matter} + H_2O + \text{Bacteria} + \text{temperature}]$$
anaerobic environment

$$\rightarrow \text{new cells} + \text{resistant organic matter} + CO_2 + CH_4 + NH_3 + H_2S$$

Within the mounds, bacteria digest organic matter to release carbon dioxide (CO_2), methane (CH_4), ammonia (NH_3), hydrogen sulfide (H_2S), and other nitrogen and sulfurous compounds. Although carbon dioxide and methane (which can be used as biofuel or burnt for carbon credits) constitute most of the gases released, there is always some amount of toxic gas or reactions that happen outside the parameters of this template. As Mala emphasized, this reaction is akin to a diagram: it simplifies the varied and complex reactions that unfold over different phases in the lifetime of a dumping ground. What needs to be remembered is that none of the dumping grounds in the city were engineered and unsegregated trash (including hazardous waste) was directly deposited into marshland. Thus, these grounds are sites where waste matter is reacting with its surroundings. Moreover, most of the separation and transformation happens through informal networks of waste workers who are on site. Thus, each dumping ground has its own unique set of conditions. As Mala said,

> Each case is different. Each dumping ground has its own complications and its own composition. So, we can control in the sense that we can have a general idea . . . and only if everything has been scientifically managed from the start. But very few dumping grounds are like that. Each dumping ground has a unique chemical, physical, and ecological makeup.

The chemical reactions of the Malad dumping ground, which started as an open landfill in marshland, happened in a particular context (Sahu 2007). Since the dumping ground wasn't on an engineered site, there were a whole host of other factors such as the interaction between organic matter and the marsh, between chemicals in the garbage, and the sewage in the creek water. These factors were also affected by topography and the height and composition of the fill. As the garbage was dumped directly on site, the leachate oozed into the marshy ground and trickled into cracks in the rock-bed. The mound was capped with layers of plastic, which cut off oxygen supply, and the construction above added further pressure. While taking me through the chemical reactions, Mala remarked that

> there is no zero-harm dumping ground. There are always gases released and leachate that needs to be managed. The question is: how much harm? Even in a well-managed dumping ground, till the reactions settle, people should not remain in the area for long periods of time. In places like Malad, the office-going and residential population could suffer long-term effects of continuous exposure to toxic gas.

Dr. Mala's response articulates the problem of solid waste management in megacities like Mumbai, the complexities involved in turning dumping grounds to land, and the ways in which ecologies are engineered in human terms. In each phase of the ground's life, the levels of emanations are measured, and over a period of time, they are supposed to come down to safe standards that are set in relation to an adult human with a healthy immune system.[10] Measures of harm and what the outcomes might be for nonhumans that share the coast remain outside these considerations. Waste infrastructures hold aspirational futures and promise the possibility of renewal through the destruction of trash (Ahmann 2019). However, such a utopian imagination relies on the controlled conversion of garbage that is much like a template reaction, albeit along a developmental timeline:

> marshland → dumping ground + waste-to-energy
> programs → capped and turned into parks and real estate

The problem, as Dr. Mala pointed out, is that when the two transformations—the chemical conversion of waste and the transformation of dumping grounds to saleable land—are put together, the juxtaposition is far from smooth or complete. The pressure and timeline of creating saleable land from marsh often does not match up with the time taken for reactions to complete, for gases to dissipate, and for matter to decompose. For Mala, these plans failed because they overlooked the nature of substances. As she said, everything erodes over long periods of time and these erosions too are a part of the chemical life of the dumping ground. They remain outside the template reaction of development projects, which only attend to the short term.

The sleight of hand that turns garbage into fuel into land also needs to be understood in relation to the global carbon assemblage (Whitington 2016) where one credit represents the reduction of one ton of any greenhouse gas reduced from the atmosphere. Some years ago, the Asian Development Fund began financing the scientific closure of one of Mumbai's former dumping ground against the future carbon credits it was estimated to generate. Launched in 2007, the Gorai Landfill and Gas Capture project entails capping the dumping ground and burning the trapped gases. The project burns methane (CH_4) to release carbon dioxide (CO_2), which is considered a less potent greenhouse gas. The project earns tradable certificates that can be sold or used to offset domestic emissions. The project was declared a failure because the garbage only generated a fraction of the estimated carbon credits. The calculated estimates did not consider the composition of the garbage, as the vast amounts of plastic waste in it do not generate gas. The MCGM was eventually forced to buy carbon credits from the global market to pay back the US$3 million advance it received from the bank (*Indian Express* 2012).

Despite failures such as Gorai, in 2016, when the Indian government revised its solid waste management rules (MoEF 2016), it continued to rely on the classificatory and regulatory orders of urban planning and waste-to-energy programs (Mirza 2019). The government's aggressive promotion of new technologies overlooks the physical and chemical characteristics of the existing waste matter, which in India is largely composed of wet, organic matter that does not yield much energy. As trash becomes big business, it also cuts off communities of informal waste pickers, scrap dealers, and traders from the revenue stream and these precarities multiply as private companies that specialize in waste-to-energy programs or in recycling specialized goods like electronics enter the market (Gidwani and Corwin 2017; Reddy 2015).

The idea that waste presents both a crisis and a resource is not new and is tied to long histories of waste experiments in Mumbai. The successes and failures of these technological experiments reveal the sociopolitical frameworks that scaffold infrastructure—each new experiment holds the promise of modernity and a functioning state, and failures a sign of apathy or neglect (Schwenkel 2015). For example, in the late 1800s, as germ theory displaced the idea of miasmic transmission, new concepts of contagion reshaped cities (Gandy 2008). In 1890, a civil engineer by the name of C. C. James, who was commissioned to oversee the sanitation infrastructure of the Acworth Leper Asylum, designed Mumbai's first waste-to-energy plant (Doron and Jeffrey 2018). James constructed the system so that treated waste matter from the latrines first flowed into fields and fertilized fields of grass, which was sold to the city. Later, a part of the excrement was directed into a sealed chamber to produce biogas that generated electricity. James's account of the project is a careful production of human and nonhuman continuities: calibrating the volume of water used by the residents, investigating the composition of the waste, using different materials and technologies, and studying the variety and the quality of the resulting crops (James 1906).

Around the same time, in 1885, when the Health and Sanitation Departments were struggling to manage the growing quantity of solid waste, the city officials toyed with the idea of using incinerators to manage the problem. When these experimental incinerators were found to be too inefficient (because the waste was too wet), it was decided that the garbage would be transported north for use as ballast material (Attarwala 1994). This established a pattern of dumping garbage (including construction debris) in marshlands allotted for reclamation (S. Chatterjee 2019). Waste, its changing compositions, technologies, and associated public health concerns, has not only shaped Mumbai's socioeconomics, it (in its changing forms) also shares a long, physical continuity with the city. The very spaces that are described as estuarine are as much wasteland ecologies. In the existing dumping grounds of the city, under the rubble, substances that have accumulated over decades continue to transform and work in unknown ways and chronologies.

Waste presses against the technomanagerial and regulatory models of environmental management, which regards nature as a separate nonhuman domain for resource extraction (Stamatopoulou-Robbins 2014). Apart from recategorizing the landscape, this sustainability discourse laid the foundation for large-scale infrastructure interventions (Balakrishnan 2019). This policy imagined a reterritorialized coast that contained the substances and beings within the (quite arbitrary) lines it drew, and within the categories that it placed these entities in. The policy presumes to shore the coast against the polluting substances through managerial practices that consider waste matter's potential for harm in the short term, but ignore the ways in which these substances endure and act in the long term, especially the geological time scale over which the climate crisis unfolds (Hird 2013).

Contrary to the logic of the coastal policy and the narrative of death, waste's constant presence and multiple manifestations during my time in the field show that these things are not at all separable; waste is now as much a coastal substance as other things like sand and fish—that is, it is a part of the coast's chemical, ecological, and physical makeup that determine its constitutive properties and its futures. As a *coastal substance*, waste transmutes, interacts, and becomes continuous with humans and nonhumans to determine how the coast comes into being over time. These continuities of waste and their consequences sharply render the porousness of the coastal landscape and its inhabitants, as waste matter pervades in the air, water, soil, and bodies.

For the nearshore Koli fishers of Malvani (which was just up the creek from the Malad dumping ground), waste, in its many different forms (garbage, sewage, and industrial), is inextricable from social and material life across myriad scales and temporalities. Waste showed up all the time, in more ways than one. It was most immediately visible as the physical litter that the nets brought up with the fish, the open sewers that drained into the mangroves, and the soggy garbage that piled up at the entrance of the village, waiting for the weekly municipal truck. Then there was waste at work, two, three, or several steps removed: garbage and sewage react to produce algae and jellyfish blooms, which would sometimes cover the catch. It exists as industrial pollution and in the chemical load inside the fish that was sold and made its way into bodies as food. At this shore where waste abounds, it could not be considered separate in any sense. Instead, the many substances and mixes that make waste were constantly infused as new agents that were remaking the coast. These agents and their potential for action, particularly substances that take several centuries to degrade or accumulate, confound the received categories into which matter and beings are placed because of how they act in the short term. Far from inert, waste matter acts and reacts, sometimes in imperceptible ways, such as the small, innocuous-looking plastic grains that insinuate themselves into the sediment and masquerade as sand. Waste's unpredictable pathways and forms intimate a coast of interactions and continuities taking place over disjunct timelines.

MUMBAI IS ONE OF INDIA'S BIGGEST WASTE GENERATORS. IN 2016 ABOUT 11,000 METRIC TONS OF WASTE MADE ITS WAY TO MUMBAI'S LANDFILLS (CPCB 2016). THIS DOES NOT ACCOUNT FOR THE SEVERAL TONS OF GARBAGE THAT ARE DUMPED INTO THE CREEKS.

THE CITY'S SEWAGE TREATMENT PLANTS ARE ALSO INEFFICIENT, FAILING TO REMOVE THE TOXICITY FROM THE 2.092 BILLION LITERS OF SEWAGE THAT THEY TREAT DAILY (MPCB 2019).

WASTE IS MELDED INTO THE COAST, A REALITY THAT SURFACES DRAMATICALLY FROM TIME TO TIME WHEN THE SEA SPITS HUNDREDS OF TONS OF TRASH BACK ONTO THE SHORE.

WHEN RESEARCHERS MEASURED HOW MUCH PLASTIC THERE IS
ON MUMBAI'S SHORE, THEY FOUND THAT ON ANY GIVEN DAY,
ACROSS ANY OF THE CITY'S FOUR MAJOR BEACHES, A RANDOMLY
SELECTED AREA OF ONE SQUARE METER WILL HAVE ON AVERAGE
84 PIECES OF PLASTIC (JAYASIRI ET AL. 2013).

OF THESE,

42% (35 PIECES) WOULD BE MICROPLASTICS (< 1.5 MM)–
38% (32 PIECES), MESOPLASTICS (5–20 MM)
19% (16 PIECES), MACROPLASTICS (21–100 MM), AND
1% (1 PIECE), MEGAPLASTIC (> 100 MM)

1 METER

1 METER

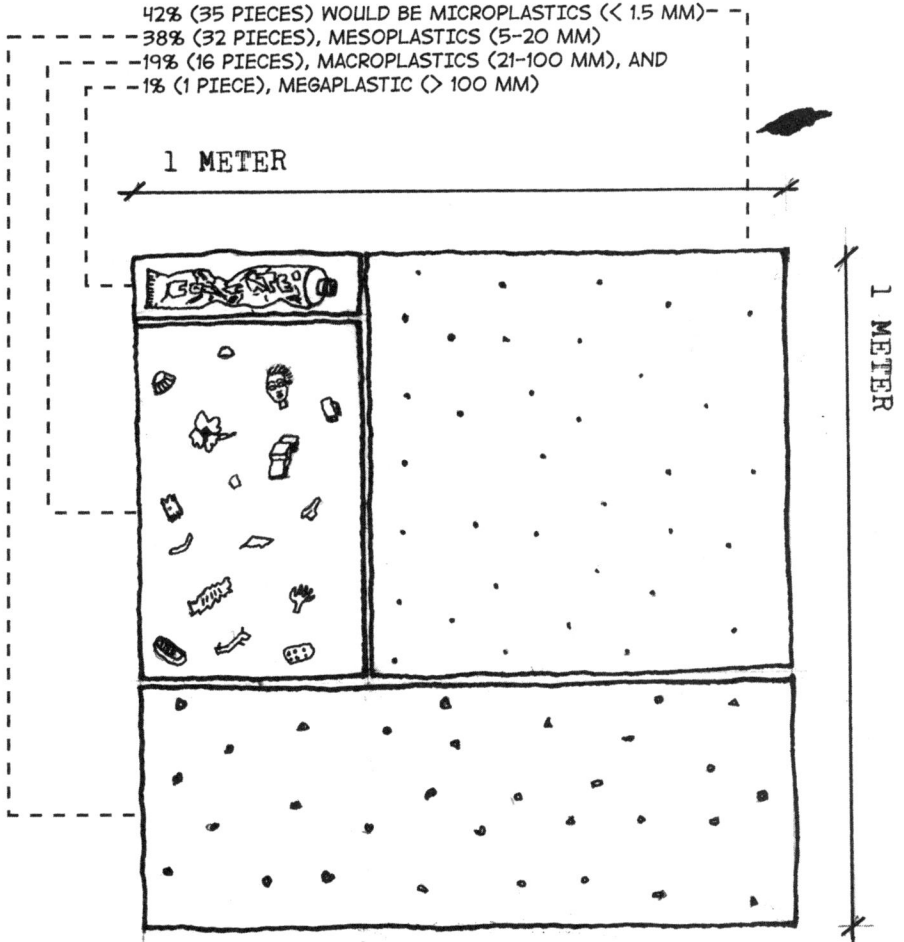

THE STUDY NOTES THAT THERE IS A LOT OF VARIATION: DEPENDING
ON THE TIME OF YEAR, THE AMOUNT OF PLASTIC CAN VARY FROM 2
TO 720 PIECES. THEY ALSO SPECIFIED THAT THE SAMPLES WERE
TAKEN ONLY ON THE SURFACE AND DO NOT ACCOUNT FOR THE
PLASTIC THAT IS INSIDE THE SAND.

Creek, Fish, Gut

Sorting involves a combination of activities apart from separating different kinds of fish. It also involves arranging the fish in containers (the best ones on top) after cleaning them in the creek. This manual labor is performed without any protective gear; bare hands dig into the mounds and separate the fish from not-fish, the edible from the inedible, and a small pile of edible-but-ugly would invariably form on the corner of the sorting mat. Unlike trawlers where quite a bit of the processing happens onboard, artisanal fishers who use bag nets find their catch in nearshore waters. They are not peripatetic, they have fixed spots allotted for each boat where they secure their nets, which take the shape of bags as they are held open by the current. The same tidal currents that trap the fish in the floating nets also carry the garbage that floats out of the city. Once in a while, I would spot boats where the entire catch was covered in garbage and black goo. Other times, our own catch and that of others would be covered with jellyfish blooms that result from pollution—the purple slime stung our hands and made our skin peel.

If the catch was completely covered with dead jellyfish, I and some of the others would be spared the task of cleaning it. While most of the catch would not be viable, cleaning it would then mostly fall on the two daily-wage laborers, Mary Bai and Jenny Bai, who worked with a few boats. Gauri Bai, who was the boat owner's wife and our boss on the beach, would yell at me:

> Aye bai! Tula zombel! Teela karu de. Uska kaam karke karke chamdi hardi ho gaya hai.
>
> [Aye woman! It will burn you! Let her do it. Her skin is tough from years of work]

And then turning to the two women, she would yell:

> Chal bai! Baghtes kai? Richav! Richav!
>
> [What are you staring at? Sort! Sort!]

Mary and Jenny would then quickly sort the catch while the rest of us worked around them, cleaning the bigger fish, hauling tubs, and running to woo the dealers. After they had finished sorting the catch, Mary and Jenny would regularly comb the beach for trash. The inedible or bad fish would be put away in plastic bags that would be sold to middlemen who sold it to fish-waste processors to produce fertilizers. Plastic and metal would be sorted and sold to the *bhangarwala* (scrap dealer); aluminum and copper items were harder to find but fetched a higher price. Mary would walk around on the beach scavenging through waste bins and under food stalls for anything that could be sold.

The infrastructural inequalities and continuities that make Indian cities have long organized relationships, disease burdens, and public health discourses in intricate ways (Ramesh 2021). In Malvani, those ties extended beyond the beach, especially because of waste. Mumbai's slums are sites where access to toilets and water is vital to everyday life as it arranges intimate relationships, alliances, associations, and ideas of community belonging; bodily comfort; and contagion (Anand 2017; Björkman 2015; Khanolkar 2021). Malvani village was surrounded by apartment blocks with private toilets. My own social and economic privilege set me apart. Everyone was aware of my upper-caste status and that I had a home with a private bath and bathroom to go to at the end of the day, just down the road. My visits to the common public toilet in the village were carefully mediated. I was warned against going to the ruined bungalow where a colonial officer had once lived; those half-broken walls were now the designated women's toilet. Few houses had a private toilet—most just had a *mori* (a shallow sink with a water trap in the floor that is usually connected to a gutter), where everything from washing clothes, dishes, and bathing, happened. Infants and toddlers also used the mori as a toilet. All the drains—including the outlets from the few houses that had a private toilet—led to open, unlined gutters that emptied out into the grounds behind the village where some families cultivated little patches of greens, which were sold in the local market. Wastewater of all sorts—sewage, kitchen waste, rainwater—was always mixing because the infrastructures that conducted these flows were porous, unsegregated, or entirely lacking, which resulted in higher health risks, especially for those without access to private toilets. The fickle arrangement of drains and toilets was exacerbated by the lack of proper water supply—the village received one hour of water every evening—which intensified the volatility of everyday infrastructure. The hodgepodge mesh of crumbling water tanks, plastic bins, metal and concrete, provisional gutters, and septic tanks had human, vegetal, animal, and microbial outcomes. While there is no systematic study that puts numbers to the rate of infectious diseases in Malvani, diarrhoea, parasitical infections, and infectious fevers were ailments that were so common they were not worthy of comment.

IN THE MID-1990S, THE MCGM INVITED TENDERS FOR ITS SLUM SANITATION SCHEME FUNDED BY THE WORLD BANK. THE SCHEME RESULTED IN LARGE FREE-STANDING TOILET BLOCKS IN A NUMBER OF INFORMAL SETTLEMENTS.

WHEN TALKING ABOUT THE SANITATION SCHEME, A FORMER MCGM OFFICIAL TOLD ME THAT THERE WERE OTHER PROPOSALS, WHICH INVOLVED FINDING WAYS TO JIGGER PERSONAL TOILETS INTO INDIVIDUAL HOMES AND HOOKING IT UP TO THE CITY'S INFRASTRUCTURE. THESE WOULD PROVIDE PEOPLE THE COMFORT AND SAFETY OF USING PRIVATE FACILITIES.

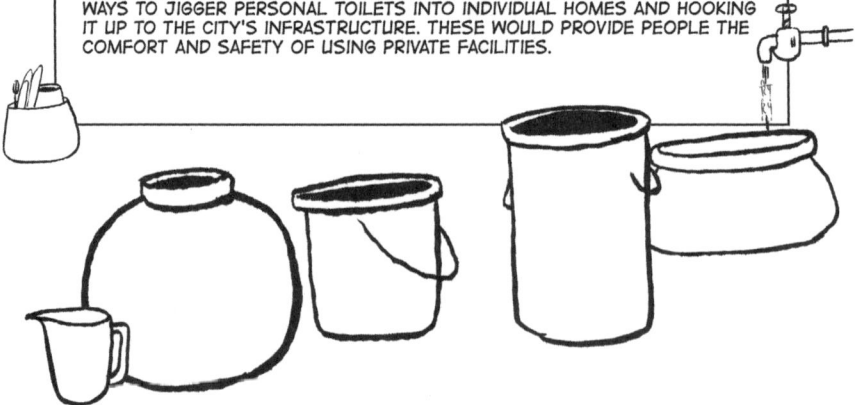

BUT PERSONALIZED FACILITIES WOULD RAISE THE COST AND, AS OUTCOMES GO, WERE LESS VISIBLE.
AND THEN THERE WAS THE FEAR THAT THE TOILETS MIGHT ALLOW PEOPLE TO CLAIM RIGHTS TO LAND OR OTHER SERVICES.
IN HIS ARTICLE ON THE SANITATION SCHEME COLIN MCFARLANE (2008) DESCRIBES THE OUTCOME OF SPARC'S SUCCESS AND ARGUES FOR MORE FLEXIBLE ARRANGEMENTS THAT TAKE THE PARTICULARITIES OF INFORMAL SETTLEMENTS INTO ACCOUNT.
THE TOILET BLOCK MODEL HAS BEEN REPRODUCED IN SEVERAL OTHER INDIAN CITIES.

The disease burdens and risks caused by the lack of sanitation infrastructure are not uniformly felt. Malvani's infrastructural ecology entangled its residents and those who worked in the fishing industry in different ways. For example, the fishing industry employs migrant workers from North India who cannot afford housing and must live on the boats. This has direct consequences on the workers' health: for those who lived on boats, the only access to fresh water or any amenity was the run-down public toilet at the beach. Leftover fish formed a big part of their daily diet, often consumed hours after they finished work at the market. Structural violence marks daily-wage laborers like Mary and Jenny Bai, who were exposed to greater risk because they are the ones doing the hard and dangerous work. Any resulting harm and distress they felt was naturalized as a result of perceived bodily differences that are often used to separate Dalit and migrant workers' bodies from others. Their skins, stomachs, and hands were deemed tougher, or as they would often say themselves when they pulled away the heavier tubs from my hands, or washed their jelly-covered limbs in the creek, that they were habituated. This perceived sensorial resilience sets workers like Mary and Jenny apart and threads them deeper into harmful continuities with waste (Gidwani 2015). Data on waste pickers in India find lack of access to protective gear, disproportionately high levels of injuries, eye and stomach infections, higher rates of psychiatric disorders, substance abuse, and finds that their poor health leads to additional economic burdens (Chokhandre et al. 2017; Salve 2020).

Waste has asymmetric socioeconomic risks and the material/corporeal outcomes of waste differ across informal communities (Gandy 2008; Butt 2020). These encounters are not just about chance meetings between individual bodies and toxic substances, but rather, they reflect the ways in which toxic materials have coconstituted environments and communities during the past century. The outcome of waste's adherences are evident in the slums close to the city's largest dumping ground in Deonar, which has been operational since 1927 (S. Chatterjee 2019). A recent baseline survey (TISS 2015) of the socioeconomic conditions of the municipal ward in which the dumping ground is located described M (East) as a ward on the edge (TISS 2015), both literally, as it is located along the eastern edge of city on the delicate estuarine landscape by the Thane Creek, and figuratively, as it has the lowest human development index compared to the rest of the city. For example, the average life expectancy in the ward is 39 years, compared to the national average of 68.5 years. The ward also performs poorly in terms of literacy, household income, access to public amenities and infrastructure, and has extremely high rates of respiratory diseases such as tuberculosis, whooping cough, and other infectious diseases—and none of these conditions can be separated from the dumping ground's presence. Even when slum communities are moved into rehabilitation schemes, the risks follow and take up different forms within these mass housing blocks (Doshi 2013). Recent studies show an increase in drug resistant strains of tuberculosis in rehabilitation schemes where apartments are stacked cheek by jowl and trash lies uncollected and festering in the alleys (Pardeshi et al. 2020).

These differentiated continuities of waste also unveil the extra-human socialities of the coastal landscape. They show how the coast is not simply the outcome of human design, rather, that it is an active landscape created in part by waste's interactions with other matter, plants, and animals. It is very hard to anticipate how different discarded substances will interact or transform over time, especially over very long periods in which inorganic matter degrades. However, research on the city's changing coast confirms that different waste substances are hard at work and remaking Mumbai's coast. For example, in the 1980s, flocks of Greater and Lesser Flamingoes started flying in to gorge on blooms of blue-green algae in the polluted marshy wetlands of Mumbai. The industrial waste, garbage, and sewage dumped into the creeks resulted in algal blooms, which are a good food source for the birds, but harmful to other creatures (Verma et al. 2004). From annual sightings of a handful of birds, the flamingoes now number in the thousands—the result of birds, effluents, and algae coming together. This confluence resulted in the death of some species and the flourishing of others. Humans too were pulled into these flamingo worlds: the birds support a flourishing flamingo-tour and bird-watching industry. In 2015, the Thane Creek, one of the sites where the flocks congregate, was turned into a wildlife reserve. In an odd, yet perhaps unsurprising turn of events, the court ordered a stay on Kanjurmarg dumping ground's expansion because it posed a threat to the flamingo habitat.

These coastal reconfigurations abound—even as the Maharashtra government celebrates its program to protect mangroves, studies show that there is a huge drop in species diversity. Much of Mumbai's mangroves now consist of just one species, Avicennia Marina, which has a high tolerance for pollution and salinity (Vijay et al. 2005). Across the coast, waste substances are disturbing existing arrangements and producing unintended coastal landscapes, which affect human lives. As fish migrate or their numbers decline with rising plankton, they portend changed patterns of existence for coastal fishing communities (Roxy et al. 2016).

Waste streams into the landscape, escaping its planned life, creating spaces in which humans and animals are pulled into configurations that have violent effects (Doherty 2019). Things get murkier when considered in the long term: even as we learn about nanoplastics in the environment and the ways in which they permeate our air, water, and bodies, we do not completely know their long-term effects (da Costa et al. 2016). Scientists researching plastic as sediment find that its effects vary across near and far geological futures as plastic mixed in sand changes the nature of sandy soil (Gabbott et al. 2020). Substances in the marsh, mud, and the creek are slowly unraveling existing relationships that sustain particular lifeforms even as they create new landscapes and ecologies (van Dooren 2014).

THOUGH *AVICENNIA MARINA* (SHRUBS PICTURED BELOW, AT LEFT) IS THE DOMINANT SPECIES IN MUMBAI'S MANGROVES, ITS FUTURE REMAINS UNCERTAIN. THE MUMBAI OF 2050 IS SET TO LOOK MUCH LIKE THE ARCHIPELAGO OF THE 1700S BECAUSE, AS THE GLOBE WARMS, THE RISING WATER WILL EVENTUALLY COVER THE LAND THAT WAS FILLED UP TO MAKE THE CITY. RANGING FROM 15 TO 20 CM IN HEIGHT, *AVICENNIA MARINA'S* ROOTS ARE NOT TALL. CONSEQUENTLY, THIS SPECIES IS NOT EXPECTED TO SURVIVE THE RISING WATER LEVELS. IN SOME MANGROVES, OTHER SPECIES THAT HAVE TALLER ROOTS ARE SLOWLY REPLACING *AVICENNIA MARINA*, SUCH AS THE *SONNERATIA ALBA* (PICTURED BELOW, AT RIGHT) (B. CHATTERJEE 2019B)

WASTE'S CONTINUITIES GO FAR BEYOND THE CITY'S MARSHES AND DUMPING GROUNDS. IN THE WINTER OF 2016, THE WAVES GLOWED BLUE ALONG MUMBAI'S SHORE, BRINGING MANY OF THE CITY'S RESIDENTS OUT TO THE BEACH AT NIGHT TO SEE THE SPECTACLE.

THE BIOLUMINESCENCE WAS LATER ATTRIBUTED TO BLOOMS OF NOCTILUCA ALGAE, WHICH PRODUCE DEAD ZONES, OR LOW-OXYGEN AREAS OF THE SEA THAT CANNOT SUSTAIN MARINE LIFE.

CLICK!

THE PATCH OF SEA BETWEEN THE GULF OF OMAN AND MUMBAI IS OFTEN DESCRIBED AS A DEAD ZONE CREATED BY THE SEWAGE, GARBAGE, AND INDUSTRIAL AND AGRICULTURAL RUNOFF THAT URBAN AGGLOMERATIONS LIKE KARACHI AND MUMBAI RELEASE INTO THE SEA.

AYE GHOCHU! FLASH NAHIN CHALEGA!
[HEY IDIOT! THE FLASH WON'T WORK!]

The Contest over Dead Zones

Mumbai's coast has some of the highest amounts of marine litter and pollution in the world, and is often called a dead zone. The amount of garbage that Mumbai produces has more than doubled in the last decade. In a very short span of time, the addition of these chemical and material agents has radically altered coastal conditions. The coastal landscape consists of diverse species of plastic of different varieties, sizes, and densities that press on plant and animal life through different kinds of encounters (Jayasiri et al. 2013). The outcomes of these encounters are not singular events, but protracted and transformative in ways that unravel the existing threads of entanglement. These encounters are as much about life transforming through the changing relationships between coastal creatures and trash as they are about destruction and death.

In 2017, I heard of a volunteer-run group called the Marine Life of Mumbai (MLOM) that was organizing public walks along Mumbai's shoreline. It was led by a small team with varied expertise in different industries and areas of research, including zoology and marine biology. The group was also invested in promoting coastal conservation and raising awareness regarding the environmental impact of developmental projects. These walks were a way for the human residents to encounter the marine residents of city, and through that moment of encounter establish a multispecies relationship that produced a different understanding of the coast—one that was not predicated on human needs and developmental desires.

As we walked along the shore, the organizers would point out various species. As he skipped over rocks and squatted over shallow pools, Sagar—one of the organizers—would stick his hand in the water and lift rocks to show us the crabs and shrimp underneath as Dilip, another organizer who was better at identifying species, would tell us the creatures' names and where they lived and what they did on the coast. They would urge participants to take photographs and post them on the group's social media page, to go on walks by themselves and share what they saw and to build a collective visual archive of life along the shore. In the distance, I could see the wastewater pipes emptying into the mudflats and I was terribly conscious of my sandaled feet soaked in gutter water. I asked Sagar, a bit tentatively, whether he was worried about being in the water, to which he replied that he had been walking up and down the coast for very long and nothing had happened to him. Then he quickly plunged his hand elbow deep into a puddle and pulled out a rock partially covered with coral and sponges. Pointing to one part of the coral that was bleached white he said, "Look this part is dead. But this other part—and the sponge—they are alive! What does that say then?"

This question about what it meant that these creatures were still alive stayed with me long after the walk was over. I wondered whether such a pronouncement would have the effect of making it seem as though the pollution was somehow acceptable or under control. During the next walk, apart from learning about sponges and jellyfish, I also gathered that some of the group members were going to a meeting to discuss the environmental impact of the Maharashtra government's plan to build a statue of Shivaji, which would be the tallest in the world, on an artificial island out in the sea. Shivaji was a seventeenth-century ruler and is considered the founder of the Maratha Empire; his mythological status has fundamentally shaped regional politics. One of the organizers, Gaurav, was talking about the ecological impact of the project on the coast. As I gathered from the conversation, in many ways, MLOM was formed as a response to unchecked coastal development, and understood in this context, walking the coast to see signs of teeming aquatic life takes on new meaning. As he was describing the upcoming meeting, Gaurav repeated what Sagar had said before: that the campaign had to show how the coast was not a dead space that could be exploited for development. There was life that needed to be conserved. He went on to explain that a number of projects had been justified and given the green light on the basis that Mumbai was an already developed urban area and therefore not worthy of conserving as there was nothing natural about its landscape. "We need them to see that there are thriving ecosystems that support marine creatures here," he said. Thus, walking the coast meant seeing the ways in which it was not devoid of life. It meant acknowledging the creatures at the shore as the city's residents. Rather instead of thinking of the creatures as occupying coastal space designed for human needs, walking was a way of inhabiting their extra-human worlds and cultivating a different understanding of the coast that was the outcome of nonhumans acting independent of/ in spite of humans. It also meant recognizing the future possibilities in relation to those creatures, recognizing the possibility of loss, and suspending one's anthropocentric view.

Today, the MLOM has become much a bigger movement.[11] The shore walks are almost always fully signed up as soon as they open. The visual archive consists of several meticulously labeled and captioned images that identify species that spring up on the coast. This archive is crucial to the group's program that the coast is not a dead zone, that it has life that needs to be seen, and that it is so much more than a site that serves urban functions. The visual encounter, which happens either on a walk or when a random visitor browses the archive online, is intended to produce an affective response. Though the images do spectacularize the underwater and intertidal worlds (Elias 2019), when seen as a collective in an archive, they do not seek to fetishize coastal space. Rather, they seek to subvert the forces that want to claim the shore for human designs.

Finding and pointing to creatures at the shore, taking their pictures and participating in the creation of community archives is an acknowledgment of entanglements and of how the coast happens through the coming together of substances and creatures. Each encounter is also a confrontation—it makes the viewer face the notion that entanglements with creatures and physical affinities with things may not always be warm and pleasant—they may hurt or induce despair, and they may be, as Sophie Chao (2022) writes, unloving. These encounters may induce discomfort, but they are important nonetheless. They make us acknowledge that danger, pain, and sadness are fundamental aspects human-animal intimacies (Govindrajan 2019).

These images are a window into the forms that these intimacies take. The creatures, and the images, are evidence of a coast that is a lively edge, slowly changing and moving with the things that make it. The fuzzy, transforming, and transmuting edges of these physical, chemical, and social intimacies reveal how Mumbai's coastal landscape exceeds the visions and timelines of planning projects. By the term *lively*, I refer to not just the "lifeforms" that inhabit the coast, but also to what Jane Bennett terms the "distinctive capacities and efficacious power of particular material configurations" (Bennett 2010, ix) that are usually not thought of as having life. Liveliness describes the capacity of this seemingly inert matter to act and to exert force on landscapes, events, other materials, and beings. This shore is made of actants and forging a relationship with a marine being and recognizing the work of waste matter allows us to see a coast that is lively and slow moving. MLOM's walks demonstrate the nonhumans at work, in order to impress on people that the coast is much more than a space for work, recreation, resource, and inhabitation for humans alone. Their images unravel the boundaries that developmental projects draw between organisms that are recognized as having life and things deemed to be dead or inert in order to justify postliberal attitudes and policies that allow the environment to be consumed, even if under the guise of sustainability. These images reveal the coast as a site of mixing, where lines are provisional—and this has important implications for the coast as an entity and object of knowledge. As Donna Haraway (Haraway 1988, 595) writes, as much as boundaries are drawn to define and distinguish, they are not set, since their "boundaries materialize in social interactions. Boundaries shift from within. What boundaries provisionally contain remains generative, productive of meanings and bodies."

These multispecies encounters—during walks and in archives—serve as good reminders that waste is an interesting substance (or mixture of substances) to think about. Heavy pollution is the reason why the Mumbai's waters are considered a dead zone—waters where the oxygen levels are depleted to the extent that it can no longer support life. This happens because waste discharge produces a profusion of algae that uses up all the dissolved oxygen. This idea of death also drives a number of urban interventions, which are justified on the basis that there is really no "nature" left to protect along the city's shore. However, the ways in which waste constantly interacts and coproduces ecologies at the edge presses against the idea of a dead zone and vivifies it in different ways. By this, I do not mean that the coast suddenly gains life—but that it confronts its human residents with the possibility of sharing it with nonhumans. Even, or rather, especially, toxic ones.

The marine life at the shore exists in intimate relationships with infrastructure and waste at the shore. These beings show that far from being self-contained, infrastructural projects serve as catalysts for interactions where organic and inorganic matter react with the soil, water, air, organisms, machines, and bodies over timescales that range between seconds, months, and years to produce physical and ecological changes. These physical, chemical, biological, and ecological reactions that take place when different kinds of matter interact often stymie and escape the controlled cartographic transformations of urban development and its planned timelines. These disparate timelines; the interstitial spaces between the planned, the unplanned, and the unforeseen; the reactions between the landscape, the elements, matter, and organisms make apparent the actions and temporal horizons of nonhumans in making the coast—both in the present and over longer timescales that are hard to comprehend yet necessary to understand.

THE FIRST TIME I WENT ON A WALK WITH MLOM, I WAS LATE. THE SMALL GROUP WAS ALREADY OUT ON THE SHORE NEAR HAJI ALI. WHEN I CLIMBED DOWN TO THE ROCKY SHORE, A PACK OF DOGS SURROUNDED ME. SAGAR SHOUTED, "THEY WON'T DO ANYTHING! JUST WALK OVER!" WHEN I REACHED HIM, HE REMARKED, "WELL, YOU WERE IN THEIR SPACE, AFTER ALL."

WE SAW SPONGES AND CORALS HOLDING TIGHTLY ONTO ROCKS. EACH SHALLOW TIDE POOL WAS THE SETTING OF A STORY: TWO CRABS UNDER A ROCK, AND LOOK "THERE'S AN EEL. DO YOU SEE THAT SNAIL NEXT TO IT? THOSE ARE BARNACLES. YES, THEY ARE ALIVE . . ."

I SAW A BLUE BUTTON JELLYFISH, A DELICATE THING SUSPENDED IN THE WATER WITH A WHITE CENTER AND BRIGHT BLUE TENDRILS REACHING OUT. AGAINST THE GRAY, BROWN, BLACK, AND WHITE OF THE ROCKS, IT HUNG, A SHOCK OF COLOR.

EACH IMAGE IN MLOM'S GROWING ARCHIVE IS A WONDROUS AND PECULIAR ENCOUNTER.

A COUPLE OF CRABS CAMOUFLAGE THEMSELVES IN THE ROCKS AT A SHORE WHERE LOVERS GO TO HIDE AND MEET IN SECRET.

STARFISH WITH LONG, SPINDLY TENTACLES COME TO THE SHORE AND ARE ILLUMINATED BY A WALKER'S TORCHLIGHT.

CORALS REACH INTO THE HUMID BREEZE.

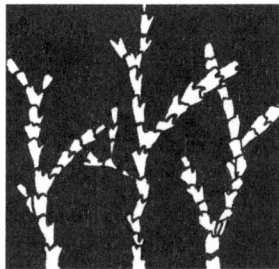

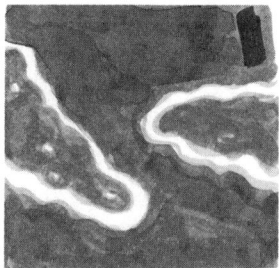

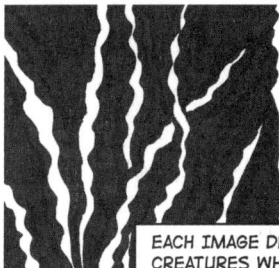

EACH IMAGE DESCRIBES CREATURES WHO LIVE AND DIE AT THIS SHORE JUST AS MUMBAI'S MILLIONS OF HUMANS DO.

THESE ENCOUNTERS MAKE IT HARD TO DENY THE MULTISPECIES AND MULTIMATERIAL INTIMACIES THAT MAKE THE CITY AND ITS SHORE.

Coastal Substance

In late 2011, Mumbai passed an ordinance banning the use of thin plastic bags, citing coastal pollution as one of the main reasons. While this was a necessary decision that was met quite favorably, it all but halted the retail business at the beach because patrons would refuse to buy fish unless it was wrapped in a plastic bag. For boats like the one I worked with, retail business formed a small part of the income they earned from fishing, but it was not insubstantial. This meant scrambling to clean bags salvaged from the beach or hiding plastic bags under the tubs in order to escape fines. The fact that I could, when needed, find a plastic bag by walking a few meters along the beach pressed against the ways in which policy initiatives seemed to overlook the condition of the shore—that it was a site where things were mixing and that waste was already a physical part of its mix.

Plastic is an interesting material to think about, given that the availability of cheap packaging is fast changing the composition of urban waste in India. As plastic proliferates, it introduces new ideas of contagion, separation, and pollution (Pathak 2020). At the beach, the ubiquitous plastic bag was a sign of how discarded things were mixing with the landscape. After the short spans of development projects and policies, waste would exist as a substance that makes up the beach, waters, fish, and coastal communities and economies—organizing life across scales toward uncertain ends.

Waste as a coastal substance departs from the management logics of current environmental policies and human designs. At the same time, I wish to underscore that such an understanding of waste does not imply that polices like plastic bans or technologies like sewage treatment plants are futile. Nor do I wish to repeat the refrain bemoaning the failing and dysfunctional infrastructure that characterizes cities like Mumbai.[12] Instead, the understanding I propose is a way of placing infrastructures installed to combat the harmful effects of human-generated waste in relation to the material qualities of the substances they deal with and seeing them as immersed in landscapes made of other materials and beings. Second, recognizing waste as a coastal substance is not a turning away from important questions about land rights, the right to work, and inhabiting a clean and healthy environment. Rather, seeing how landscapes are immersed in various temporal registers creates openings for rethinking the time frames that environmental rights and citizenship work in.

To articulate waste as a coastal substance requires breaking with nature/culture dichotomies and attending to the sociomaterial entanglements between matter, beings, and landscape. Waste materials have a range of capacities that can, in varying degrees, shape other entities—including humans, alerting us to our permeability (Hird 2013). Their actions and adherences make bodies and landscapes, sometimes in ways that have harmful effects across time (Alaimo 2010). Thinking in terms of how things act allows us to move away from the idea that landscapes are products of human design and from the idea of the human body as self-contained and separate from the environment. When seen in terms of things, their actions, and their continuities, the coast is not just the outcome of human development, but a landscape that makes itself through an emergent collection of entities acting in time. The humans, microbes, animals, plants, landscape, and things that make the coast are not necessarily discrete; they share a range of changing connections.

Thinking with things as agents that have changing capacities also reveals the paradox that lies at the heart of environmental management—particularly waste management. Mumbai's waste infrastructure treats the coast as a zone or space where waste belongs and where it can remain separated until it is classified as something safe or useful for humans. Marshy estuarine areas are thus regarded as rightful sites for waste, which once deposited, becomes productive material that will eventually turn into land. Mumbai's coast is also the site where sewage and industrial run-off are released in the nearshore waters and at a certain dilution, they transform from waste into the sea. These transformations happen through various kinds of mixing, such as the interactions between waste and waste workers, between waste and marsh, sewage and seawater, and they continue well after development projects are completed. Thus, even as the coast is set apart from waste as a polluting substance, it is paradoxically a site where waste interacts with manifold entities and where substances and beings constantly escape categories, space, and time.

Whereas large-scale projects deal with the immediately visible problem (the garbage that is there), they miss the ways in which slow and unsighted violence (Nixon 2011) happens through the imperceptible yet persistent action of things. As waste matter mixes and meets other materials and beings, its components may work immediately, and at other times, their action may be deferred. Waste introduces a plethora of latencies (Murphy 2013), which describe the possible outcomes that may result from interactions in the present. These latencies show how things are not just meeting, mixing, or becoming continuous in different ways (which they are), but that continuities happen in time and can produce complex, unanticipated ecologies, especially over very long periods (Hoag et al. 2018). Thus, thinking of waste as a coastal substance shows how the coast is not just a space, but is an entity immersed in time through interactions that cannot be predetermined and yet have important consequences for human lives.

Landscapes are assembled through the actions and material qualities of nonhumans that unfold in the deep long term. The varying effects of substances, especially their harmful outcomes, alert us to the ways in which the humans, animals, and plants are open ecosystems, and for the capacity of things to penetrate and cohere with their bodies, even if those cohesions are unwelcome. These toxic continuities seep into human lives at different timescales. There is the quick encounter with a caustic agent or the sudden harmful accumulation of poisons—such as oil spills. Then there are the continuities that produce effects over a long period of time, which sometimes make it harder to mitigate harm or to point to discrete causes. They work slowly, in ways that obfuscate their scale of destruction such as the changing composition of marine life at the shore. The narratives that scientists, marine life advocates, and fishers articulate are not answers to what might happen in this long-term future, but they do offer hints and points of entry to perceiving where human bodies, creatures, matter, and things that happen in the short-term lead into the long term. Far from straightforward, the ruptures, gaps, and buried elements of these narratives describe a coast that slowly changes and moves with the things that make it.

DRAWING COASTLINES

"It's a good place to land," the foundress said in her sweet soprano, examining the first rough maps that the scouts brought back. There were plenty of caterpillars, oaks for ink galls, fruiting brambles, and no signs of other wasps. A colony of bees had hived in a split oak two miles away. "Once we are established we will, of course, send a delegation to collect tribute.

"We will not make the same mistakes as before. Ours is a race of explorers and scientists, cartographers and philosophers, and to rest and grow slothful is to die. Once we are established here, we will expand."

—E. Lily Yu, "The Cartographer Wasps and the Anarchist Bees"

IN E. LILY YU'S (2011) STORY, THE BEES ARE DOOMED EVEN BEFORE THE WASPS ARRIVE. THE WASPS' CONQUEST BEGINS WITH A SKETCH THAT PREORDAINS THE FUTURES OF HIVES, TREES, MOTHS, AND GENERATIONS OF BEES.

THE WASPS TRAIN THE BEES IN THE VISUAL SCIENCES OF COLONIZATION, TO DRAW IN ELABORATE DETAIL, THE ACT OF POSSESSING A LANDSCAPE.

THESE DRAWINGS TURN BEE WORLDS INTO WASP WORLDS.

In 2019, the Bombay High Court ordered a stay on the Mumbai Coastal Road Project—a nearly thirty-kilometer-long, eight-lane expressway under construction along the city's edge. When completed, the road will connect the southern end of the narrow, triangular city with the western suburbs to the north. The road will cost US$1.7 billion, and though the expressway itself only requires twenty hectares of land, the project will end up reclaiming ninety hectares, displacing and destroying coastal communities and drastically altering the physical makeup of the shore. The city's planning authorities have pressed on with the project despite numerous court injunctions and protests from fisher communities and environmental activists. In 2019, this motley resistance group won a short-lived victory against this massive project. The reason behind this victory was a marine survey, which had found the presence of coral colonies along stretches earmarked for reclamation. Since corals are a protected species under the Wildlife (Protection) Act passed in 1972, the court ordered a halt to the project. And thus, a fragile community of nonhumans on a marshy edge was able to pause a massive project that draws a sharp line between the land and sea.

When the project was first proposed, the committee that studied its feasibility listed "congestion" as the main reason for creating the transport corridor. The committee supported it on the grounds that Mumbai's roads were too congested due to the sharp rise in vehicular traffic ("Coastal Road Mumbai" 2011). The report pointed out that between 2004 and 2011, the number of vehicles on Mumbai's roads rose from 12.3 million to 19.1 million, an annual increase of 96,000 vehicles. Congestion also had another dimension: the problem was not only traffic density, but also vehicular speed. The committee argued that the congestion in the city had lowered the average speed by eight kilometers an hour in some areas, which produced more vehicular emissions. The city needed the additional road to raise the speed closer to ninety kilometers an hour, as cars operated more efficiently at that speed, thus reducing emissions (though the report does not provide a source for this claim or contextualize it). Other justifications included the lack of public promenades, jogging tracks, and beautifying the seafront: "While the long coastline of the city is its asset, the sea fronts in the city are either abused by unsightly encroachers or happen to be private backyards of the chosen few and not available to the public unlike beautiful waterfronts and gardens/promenades adjoining the sea in major coastal cities of the world" ("Coastal Road Mumbai" 2011, 3).

In the 3D visualizations published in newspapers, the Coastal Road is a swirling mix of grey lines and neat green fields that cut a well-defined edge against the sea. Data on public transport use contradict the simple and neat picture that the proposals and visualizations present—approximately 70 percent of the city's population travels by public transport, walks, or cycles. Less than

20 percent of the city uses cars (closer to 10 percent) or two-wheelers (about 8 percent) (Center for Science and Environment 2017). The project seems even more implausible considering the damage it has inflicted in the few years since it began. It has damaged intertidal habitats and the destruction caused by it has hit the human and nonhuman communities of the coast particularly hard. The noise and disturbance from the project have changed the movement of the fish. The physical changes to the coast have destroyed natural barriers that protect the boats from the force of waves. The Koli community has resisted the project at every turn—often taking their boats to occupy the sea and disrupt construction and reclamation activities. Yet, despite the numerous protests and reports that raise environmental concerns, the Maharashtra government continues to support this project. Therefore, the 2019 verdict seemed like a significant victory and a long-sought acknowledgment of the city's more-than-human coastal worlds. It disrupted—for a while at least—landward imaginaries of the city in favor of seaward visions (Siriwardane-de Zoysa and Hornidge 2022).

Though the courts halted the project, it did not pause for long. Shortly after the High Court order came through, the Supreme Court ruled that the Coastal Road would continue as it was in the public's interest. It sided with the state's argument that the population of corals was too small to be used as a justification against a major transport corridor. It greenlit a plan to relocate the coral colonies to other areas along the shore (Bhutia 2020).

The coral colonies' fate mirrors that of residents of informal settlements relocated under the Slum Rehabilitation Act (1995). Just like these residents, the corals too were enumerated and documented. Just like the social, economic, and health risks that relocated residents face, the coral relocation project too poses risks. It was not clear whether this project would be successful. The project posed it only as a question of survival—whether the grafted coral would take hold. However, what of their rich worlds? Despite these doubts, shortly after permissions came through, teams of scientists and their assistants appeared with buckets, hammers, and chisels. They chipped away at the rocks to dislodge the coral and move them to their new homes (Bhutia 2020).

These events unfolded during the COVID-19 pandemic that made travel to Mumbai, my hometown, impossible. The story of the corals affected me deeply because I remembered the wonder of encountering them on my walks with Marine Life of Mumbai (MLOM). I recalled jumping off the Haji Ali causeway, walking across the marshy, rocky shore as the organizers turned stone after stone and showed us corals, algae, and other marine creatures in tide pools. I realized that perhaps such encounters would not be possible at that edge. However, I hope that the corals take hold where they go and that they thrive.

This account illustrates the dilemma that confronts the narratives in this book and environmental action more generally: that of the persistence of anthropocentric plans and developmental timelines despite evidence of the long-term damage they cause and their asymmetric, violent outcomes. The visual efforts that fishers, urban activists, and marine life advocates perform does work against these conceptual and temporal frameworks. However, they do not have a triumphant outcome, much less a clear resolution. In the decade that has passed since the 2011 CRZ policy, the fundamental orders of property and resource continue to guide environmental governance. Subsequent coastal policies have systematically rolled back environmental oversight. Rainfall at the scale of the 2005 flood is no longer uncommon, and the frequency with which cyclones hit the city is alarming. The visual projects in this book offer entry points for thinking about climate action that reach beyond the language of rights and resilience—but they are also plagued by the lack of resolution, the mitigating conditions, and disappointing turn of events.

This book attends to the problems of anthropocentrism and solution-oriented developmental horizons through the act of drawing, which it frames as a process that introduces doubt, renders multispecies relationships, and gestures toward other temporal frameworks. It explores technical drawing practices and pays attention to how they infuse objectivity into the image. It draws a line that connects these scientific images with extractive orders. This connection is a result of the models and conceptual frameworks that inform drawing practices as well as their technical instruments, mediums, and modes of circulation.

The first part of this book shows how the outcomes of this connection are evident in recent coastal changes. By 2011, the coastal policy shifted from a view that divided the coast in broad swathes to a more granular one. The development of satellite technologies is central to this shift. Higher-resolution imagery allowed policymakers to exercise greater control over smaller sections of the coast. The availability and ease of accessing geographic information further tightened the relationship between technical accuracy and authority. However, as the first chapter shows, accuracy and associated authority are not the automatic outcomes of technological development—they are produced through drawing practice and through the complex relationships between creators, instruments, and elements in the landscape.

The new policies recast coastal subjects in ways that mirror the control and appropriation of landscapes: small-scale fishers are recast as traditional fisher communities, as both wardens of nearshore ecologies and as vulnerable subjects. The residents of informal settlements are turned into vulnerable subjects because of the material conditions of the settlements. Even though the effects of coastal erosion, flooding, and sea-level rise introduce greater risks for some communities than others, the ways in which vulnerability is framed, understood, and responded to is often problematic (Marino 2015). These technical instruments and policies turn coastal nonhumans into objects—they reduce them to "mere ground and matrix" for developmental projects (Haraway 1992, 11). Simultaneously, they also elide coastal communities whose relationships to the shore do not fit within the neat resource-oriented categories.

If, in the beginning, this book attends to the practices that construct ideas of objectivity and accuracy, it then turns to look at the transformations that come about as a result of these drawings. However, as I believe I have shown, the lines in these drawings take discursive forms: in the fisheries industries, they become the basis for calculating the extractive potential of the coast and establishing economic goalposts. They turn a fluid coast into one that is increasingly terranean. The lines of graphs that trace extreme rainfall become the basis for turning meandering and hybrid channels into effective passages for carrying rainwater to control urban floods.

The violent outcomes of these extractive economies and of climate disaster set up the book's central provocations: where do the possibilities of drawing different kinds of coastal drawings reside? How can drawings manifest qualities from whence a sense of texture, memory, and experience of landscapes emerge? How can critiques of human intervention go beyond merely pointing to entanglement, and how can it, instead, find tactics for political action (Giraud 2019)?

To get at these questions, the book focuses on particular moments, practices, communities, and drawings. First, it looks at moments of doubt that infiltrate "accurate" drawings: it looks to nonhumans, such as the plants that mark the tide line, or marine creatures that persist in polluted waters and resist representational reductions. It is attentive to places where instruments and inscriptions introduce doubt, and where fluidity impresses on coastal lines and introduces depth into a landscape simplified to a flat plane. The process of drawing accurate pictures also reveals the moments that introduce doubt and the possibility for negotiations in a process that, on the face of it, seems foreclosed. Second, it looks at practices such as small-scale fishing, which, when released from the confines of the "traditional," becomes a means of attending to coastal complexity, fluidity, spatial depth, and temporal depth. The socioeconomic webs of nearshore fishing offer ways of drawing and designing more environmentally sound and equitable fisheries. The interactions and intimacies of nonhumans and the ways in which they resist becoming simply a resource displace the anthropocentrism of development plans and reveal ways of living geologic time. In the harmful actions of waste and toxic chemicals, one finds the possibility of building plans that are freed from the ten-, twenty-, fifty-year cycles of urban projects and ask us to consider what a coast might be thousands of years from now. New digital archives reveal toxic intimacies, the deferred action of pollutants, and the asymmetries of life in a polluted world. Simultaneously, they also reveal the possibilities of drawing as political action. They offer conceptual openings for anthropology: to think about how lines may convey time, how granularity might enter ethnographic narratives of coasts, how flat planes become deep, and how to see the shifting lines between land and water over time.

Rather than thinking of scientific images as antithetical to these quests or resorting to simplified narratives of resistance and counterimage making, the visual projects in this book lean toward the productive possibilities of technical abstractions that embed human communities in coastal worlds and timescales. I suggest that one way of coming at the unsurmountable climate predicament that laps at the shore is to draw more—particularly technical images—but draw them differently. The examples in this book do not provide solutions for exactly what those different drawings look like—but, I hope, they do offer important conceptual starting points for drawing, making different coasts, and a case for anthropology as a good source of those starting points. They make a case for turning toward living and planning for a polluted shore where the water's edge is slowly, but surely, moving in. They suggest crafting continuities between human-animal-plant-matter-landscape through new ways of drawing. Drawing in ways that recognize how the coast comes alive (and dies) through the momentary reactions of volatile chemicals that are dumped into the water, the timelines of urban planning, through seasons and life cycles of fish, through events of extreme rainfall and flood, through formations of littoral flora and fauna, and through deep horizons of ecological damage, toxicity, and the (im)possibility of recovery.

Notes and Credits for the Comics

Introduction

p. 21 Nikhil Anand and Caroline Terens. 2019. "Explained: How Climate Change Could Impact Mumbai by 2050," *Indian Express*, December 13, Mumbai edition, sec. Explained. https://indianexpress.com/article/explained/the-rising-threat-to-mumbai-6160595/#:~:text=The%20research%2C%20carried%20by%20many,city%20has%20been%20built%20on. Panel images sourced from Climate Central's *Coastal Risk Screening Tool*. Online application available at https://coastal.climatecentral.org/map view/11/432.8282/19.1207/3bb40ca176d11a1c6181084d67fac7119c2b49d4b082bb45b e47ef24c7cbf6a6.

p. 22 Bombay City, Bombay Suburban, and Thane Districts. Published under the direction of Lt. Col. C. P. Gunter. 1926. Sourced with permission from the Maharashtra State Archives. Copy of a map of the Islands of Bombay and Colaba prepared for Mr. Murphy. 1843. In *Materials Towards a Statistical Account of the Town and Island of Bombay: Vol. 3. 1894* (Government Central Press, Bombay). In the public domain. Available at https://archive.org/details/in.ernet.dli.2015.280873. Sharada Dwivedi and Rahul Mehrotra. 1995. *Bombay: The Cities Within* (Mumbai: India Book House). Gyan Prakash. 2010. *Mumbai Fables*. Princeton, NJ: Princeton University Press.

p. 23 Vyjayanthi Rao. 2007. "Proximate Distances: The Phenomenology of Density in Mumbai," *Built Environment* 33 (2): 227–48. https://www.jstor.org/stable/23289578. Salman Rushdie. 2010. *Midnight's Children* (New York: Random House).

p. 24 Maura Finkelstein. 2018. "Ghosts in the Gallery: The Vitality of Anachronism in a Mumbai Chawl," *Anthropological Quarterly* 91 (3): 937–68. ISSN 0003-5491.

p. 29 The bottom panel references the *SOAK* exhibition held at NGMA. Anuradha Mathur and Dilip Da Cunha. 2009. *SOAK: Mumbai in an Estuary* (New Delhi: Rupa & Company).

p. 30 Davies's sketch of Bombay Harbor, 1626. Drawn from the *Gazetteer of the Bombay Presidency*. 1896. Edited by Sir James M. Campbell. General index by R. E. Enthoven. Map of Bombay and District prepared for Peshwa Madhavrao by the Peshwa's agent in Bombay, ca. 1770. Available online via the Dr. Bhau Daji Lad City Museum, Mumbai. Part of a Surat-to-Bombay map by Benard. Pierre-Marie-François Pagès. 1782. *Voyages Around the World, and to the Two Poles, by Land and by Sea, During the Years 1767, 1768, 1769, 1770, 1771, 1773, 1774 & 1776* (Paris: Chez Moutard). In the public domain. Akshay Kore. 2016. *MumbaiData Map*. Available at https://akshaykore.github.io/mumdata/. Used with permission.

Chapter 1

pp. 39–40 For coastal length, see "Shoreline Change Atlas of India (Volume—2 Maharashtra and Goa)." 2014. Atlas SAC/EPSA/GSAG/GSD/A/01/14. New Delhi: Space Applications Centre (ISRO) and Coastal Erosion Directorate, Central Water Commission, Ministry of Water Resources. Also, Lauretta Burke, Yumiko Kura, Ken Kassem, Carmen Revenga, Mark Spalding, and Don McAllister. 2001. *Pilot Analysis of Global Coastal Ecosystems* (Washington, DC: World Resources Institute). For fractals and coasts, see Thomas

M. Lekan. 2014. "Fractal Earth: Visualizing the Global Environment in the Anthropocene," *Environmental Humanities* 5 (1): 171–201. Note regarding scale: Map scale refers to the relationship between the size at which an area/object is represented and its actual dimensions. For example, on a 1:50 scale map, an area will be drawn at 1/50th its actual size. Larger-scale maps such as property surveys show more detail. Smaller-scale maps such as topographic sheets and maps in atlases show less detail, that is, they show a much larger area at a smaller size. Scales are relative: at 1:4,000, property surveys are larger-scale maps than topographic sheets drawn at 1:50,000 scale.

pp. 43–45 Einar Sneve Martinussen, Jørn Knutsen, and Timo Arnall. 2014. "Satellite Lamps," *Kairos* 19 (1). Available at https://kairos.technorhetoric.net/19.1/inventio/martinussen-et-al/index.html. The images were sourced from the video installation. Used with permission. William J. T. Mitchell. 2004. *Me++: The Cyborg Self and the Networked City* (Cambridge, MA: MIT Press).

pp. 60–62 Ministry of Environment and Forests (MoEFCC). 1991, 2011, 2018. Coastal Regulatory Zone Notification.

Chapter 2

pp. 87–92 Figures 36–41 appeared in "Drawing and Fishing a Salty Shore," *Anthropology News* website, October 15, 2021. © American Anthropological Association 2021. Available at https://www.anthropology-news.org/articles/drawing-and-fishing-a-salty-shore/.

p. 102 Data sourced from the Handbook of Fisheries Statistics. 2014. Section A: "Production And Disposal." Available at https://dahd.nic.in/related-links/handbook-fisheries-statistics-2014; and Handbook of Fisheries Statistics. 2020, 5). Available at https://dof.gov.in/sites/default/files/2021-02/Final_Book.pdf.

p. 103 FAO. 2022. *The State of World Fisheries and Aquaculture 2022. Towards Blue Transformation* (Rome: UN Food and Agricultural Organization).

Chapter 3

p. 121 Data from the Fact Finding Committee Report. 2006. "Fact Finding Committee on the Mumbai Floods." Municipal Corporation of Greater Mumbai.

p. 127 Jacqueline Urla. 1993. "Cultural Politics in an Age of Statistics: Numbers, Nations, and the Making of Basque Identity," *American Ethnologist* 20 (4): 818–43. DOI: 10.1525/ae.1993.20.4.02a00080. Fact Finding Committee Report. 2006. "Fact Finding Committee on the Mumbai Floods." Municipal Corporation of Greater Mumbai.

p. 128 Rain and tide data sourced from the Fact Finding Committee Report. 2006. "Fact Finding Committee on the Mumbai Floods." Municipal Corporation of Greater Mumbai.

p. 129 Diagram of a vortex references an infographic that was published online by *Outlook* magazine on August 15, 2005, titled "Major Flooded Areas." Before and after maps of the Mithi at the airport drawn using "Development of Action Plan for the Environmental Improvement of the Mithi River and Along Its Banks." 2006. Graph of rainfall data drawn using Jenamani et al. 2006. "Observational/Forecasting Aspects of the Meteorological Event That Caused a Record Highest Rainfall in Mumbai," *Current Science*, 1344–362.

pp. 136–37 I have created the tables and graphs using data from infographics that appeared in the *Times of India* over one monsoon season, with the following articles: Richa Pinto. 2019. "Monsoon May Arrive in Mumbai in a Day or Two." *Times of India*, June 25, sec. Mumbai News. https://timesofindia.indiatimes.com/city/mumbai/

rains-advance-may-arrive-in-city-in-day-or-2/articleshow/69932760.cms; Times News Network. 2019. "Monsoon Covers Ground: Mumbai Gets 97% of June Rain in Two Days," *Times of India*, June 30. https://timesofindia.indiatimes.com/city/mumbai/monsoon-covers-ground-mumbai-gets-97-of-june-rain-in-two-days/articleshow/70006959. cms; Neha Madaan. 2019. "New IMD Data Reveals This Is Mumbai's Wettest July in 112 Years," *Times of India*, July 30, sec. Mumbai News. https://timesofindia.indiatimes. com/city/mumbai/new-imd-data-reveals-this-is-citys-wettest-july-in-112-yrs/article-show/70441656.cms; Richa Pinto. 2019. "3,453mm & Counting: Mumbai Breaks 65-Year Rain Record," *Times of India*, September 13. The data in the articles are publicly available via the IMD.

p. 151 Aslam Saiyad, aka Bombay ka Shana's work can be found at https://www. instagram.com/bombay_ka_shana/. Name disclosed with permission.

Chapter 4

pp. 158–60 Background images by Sagar Shiriskar. Used with permission.

p. 166 Background image refers to the advertisement in *Times of India* on March 3, 2012. Available at https://pkdas.com/news/Open%20Mumbai%20ad%20 02 png.

p. 183 Percival Strip and Olivia Strip. 1944. *The Peoples of Bombay* (Bombay: Thacker).

pp. 184–85 Woman Holding a Fish on Her Head, by John Griffiths, Mumbai, India, 1872. Koli woman painted by M. V. Dhurandhar appears in Stephen Meredyth Edwardes. 1912. *By-Ways of Bombay* (Bombay: DB Taraporevala Sons), 13. In the public domain. Kajri Jain. 2007. *Gods in the Bazaar* (Durham, NC: Duke University Press). Partha Mitter. 1994. *Art and Nationalism in Colonial India, 1850–1922: Occidental Orientations* (Cambridge: Cambridge University Press).

Chapter 5

p. 200 F. N. Attarwala.1994. "Solid Waste Management in Greater Bombay: Final Report." Mumbai: National Environmental Engineering Institute. Shalini Nair. 2007. "Gases Spook Comps in IT Park Built on Dump | Mumbai News—Times of India," *Times of India*, April 2, 2007. https://timesofindia.indiatimes.com/city/mumbai/Gases-spook-comps-in-IT-park-built-on-dump/articleshow/1842135.cms. Amiya Kumar Sahu. 2007. "Present Scenario of Municipal Solid Waste (MSW) Dumping Grounds in India," in *Proceedings of the International Conference on Sustainable Solid Waste Management*, Mumbai, India, 5–7.

p. 202 Syantani Chatterjee. 2019. "The Labors of Failure: Labor, Toxicity, and Belonging in Mumbai," *International Labor and Working-Class History* 95: 49–75. DOI: 10.1017/S0147547919000073. Data for the graph sourced from Abhishek Dutta and Wanida Jinsart. 2020. "Waste Generation and Management Status in the Fast-Expanding Indian Cities: A Review," *Journal of the Air & Waste Management Association* 70 (5): 491–503. Background image was the "image of the day" on the NASA Earth Observatory website on February 2, 2016. Image by Joshua Stevens, using Landsat data from the U.S. Geological Survey. Available at https://earthobservatory.nasa.gov/images/87429/ fire-burns-in-mumbai-landfill.

p. 210 CPCB (Central Pollution Control Board). 2016. "Solid Waste Generation in 46 Metro-Cities." India. MPCB (Maharashtra Pollution Control Board). 2019. "Annual Report 2018–19."

p. 211 H. B. Jayasiri, C. S. Purushothaman, and A. Vennila. 2013. "Quantitative Analysis of Plastic Debris on Recreational Beaches in Mumbai, India," *Marine Pollution Bulletin* 77 (1): 107–12. DOI: 10.1016/j.marpolbul.2013.10.024.

p. 215 Colin McFarlane. 2008. "Sanitation in Mumbai's Informal Settlements: State, 'Slum,' and Infrastructure," *Environment and Planning A* 40 (1): 88–107. DOI: 10.1068/a3922.

p. 219 Badri Chatterjee. 2019b. "95% of Mumbai's Mangroves Could Perish from Rising Sea Levels: State," *Hindustan Times*, November 16, 2019.

Conclusion

pp. 233–34 Lily E. Yu. 2011. "The Cartographer Wasps and the Anarchist Bees," *Clarksworld Magazine* 55 (April).

Notes

INTRODUCTION: SIGHTLINES

1. The process of making technical images can take nonoptical forms (Beaulieu 2014).

2. See Mariam Dossal's (2010) book for a history of surveying and drawing the city, and Sheila Patel's (2004) work illustrates how maps and plans became important for negotiating housing rights.

3. While the images may be considered as a collective, it is not necessary that they forge a global environmental consciousness (Jasanoff 2004).

4. For example, see Kumar et al. 2014.

5. As Kapil Raj (2007) writes, cartography emerged as a scientific practice through encounters between surveyors and the landscape and thus, should not be taken as a discipline separate from or imposed on the landscape.

6. A *wada* is a settlement.

7. Mumbai's settlements are often called slums, a problematic term that marks these spaces as dysfunctional, unsafe, and unclean. I use the term "informal settlements" instead.

8. I have written more about this problem of classification in an essay on housing transformation in the city. See Venkataramani 2017.

9. Even if everyone living in the settlement knew of its history, it was hard to claim rights under the CRZ without a legal record.

10. In 2014, the Ministry of Environments and Forests (MoEF) was renamed as the Ministry of Environments, Forests, and Climate Change (MoEFCC). For the sake of consistency, I use MoEFCC throughout.

11. See also Lalitha Kamath and Marina Josephs's (2015) article on public participation in city planning. Images—housing plans, city plans, experiments in drawing and designing landscape—are important parts of each of these examples.

12. I am an alumna of KRVIA and since 2006, have worked as visiting faculty from time to time. This connection was crucial to gaining access to Malvani Koliwada. While I was a part of the urban studies studio, I was not a faculty member in any supervisory capacity. In the course of my work, I also met with faculty members at the Indian Institute of Technology, Bombay, and this was possible because I had prior connections as a former resident and alumna of the institute. These connections and privilege were key factors in my ability to conduct this research. Later, my position as a faculty member opened doors as well.

1. TERRANEAN COASTS

1. Ministry of Environment and Forests (MoEF), *Coastal Regulatory Zone Notification*, 2018.

2. Ministry of Environment and Forests (MoEF), *Coastal Regulatory Zone Notification*, 1991.

3. For instance, as Sumathi Ramaswamy (2010) writes, the patriotic sentiment that freedom fighters felt on seeing an image of Mother India was not necessarily a product of an accurate depiction, but one that was definitely connected to the cartographic geo-body

of the nation. Of cartographic scale, Edney (1997) writes that it played an important part in the visual consolidation of the empire—though large-scale maps were important for gathering revenue, it was the smaller-scale topographic maps that allowed viewers to comprehend the geo-body as one whole.

4. But the leap from representation to territorialization is a complex one (Thongchai 1994).

5. James Scott (1998) makes a point about the transformative power of the map. However, Scott locates the power of maps in the people who create it—"This transformative power resides not in the map, of course, but rather in the power possessed by those who deploy the perspective of that particular map" (Scott 1998, 87).

6. By the time I joined the survey, the staff had already covered much of the area in the southern parts and had surveyed the Island City area. I accompanied groups who surveyed the *nalas*, creeks, and beaches in the northern suburbs of Kurla, Ghatkopar, Mankhurd, Malad, and Manori among others.

7. The HTL is one of the most important factors in determining the different zones in the CRZ. Almost all zones or areas are determined and defined in relation to this line.

8. Drone technology has changed surveying in important ways, though the general principal remains the same. Surveyors who use drones for aerial surveys place high contrast markers of black and white squares on the ground and record their coordinates. They locate those markers in the pictures taken by their drones and assign coordinates to align the drawing in digital space.

9. These subjective choices are critical to producing the "objective" cartographic record (Harms 2020).

10. The authors of the Coastal Zone Management Plan of the Kochi Corporation (National Centre For Earth Science Studies 2014) note:

> The most difficult part is the transferring of information from imageries to unprojected cadastral maps on which CRZ maps are prepared. This is overcome by using sufficient number of precise reference/control points spread over the entire study area for geo-referencing and compartmentalizing the maps. One of the major contributors to errors is those occurring while reproducing the cadastral map from original map through photocopying and scanning. While photocopying the enlargement or reduction produce the scale error; also the shrinkage/folding of paper during the process. Another is the scale error during geo-referencing the map. It may be noted that cadastral maps have no projection while the images are projected.

Available online at https://kerala.gov.in/documents/10180/ed62162c-f19f-4d2b-82f4-cd57e0dc6948

11. See Campling and Colás (2018) and Mansfield (2004) for studies on how land-based systems of appropriation and control are imposed on the seas.

12. The central government also used the powers conferred to it under Article 253 of the Indian Constitution, which gave it the power to create laws for the entire country in order to implement international treaties and agreements such as the 1972 United Nations Conference on the Human Environment in Stockholm, which eventually led to the creation of the United Nations Environment Programme.

13. See also Dwarakish et al. 2008; and Nayak 2004.

14. Minutes of the thirty-first meeting of the NCZMA held on May 24, 2016. Available online at http://moef.gov.in/wp-content/uploads/2017/08/NCZMA.pdf.

15. AIR 1996 SC 1446.

16. Noted in the amendment published by the MoEFCC in 1994.

17. Nayak (2002) notes that in the 1990s, data from the IRS series could be used to produce base maps at the topographic scale to show tidal action and formations such as coral reefs. A decade later, with the availability of high-resolution data from newer satellites, state institutions could revise these maps to prepare plans at larger scales.

18. At a scale of 1:12,500, 8 cm equals 1km. Therefore, a line that was 1 mm wide would in fact become 12.5 meters wide if one zoomed in to look at property level demarcations.

19. Cadastral surveys document boundaries of individual properties and their use. In India, cadastral survey sheets are generally plotted at 1:4,000 scale.

20. Also see the judgment in *Institute Of Social Welfare vs. State Of Kerala* (WP(C).No. 1050 of 2010(S)) on February 11, 2011.

21. In any kind of boundary dispute, maps that are authorized by the state carry much greater weight than independent surveys. Lee (2005) gives an account of the relevance of the form, content, and source of maps in international tribunals and frontier disputes.

22. As Wainwright and Bryan's (2009) study of indigenous counter-mapping in Nicaragua and Belize shows, independent mapping projects have to bear the marks of contracts, reports, and studies in order for those maps to gain legal recognition.

23. Hull (2012) writes that the ways in which documents reference each other is a crucial aspect of constructing bureaucratic authority.

24. Source http://www.environment.tn.nic.in/images/maps/Kan%20map/approved/app4.jpg.

2. TIDE LINES AND LIVES

1. It is important to note that there are different kinds of fishing, including inland, marine, and aquaculture. For the purposes of this chapter, I am focusing on marine fishing that happens relatively close to the shore.

2. This is partly a result of linking area with catch potential.

3. The second blue revolution proposes to increase fisheries revenue by growing the fish farming industry. This is different from the first blue revolution where the focus was on modernizing craft and moving fisher communities toward more high-yield practices (Bavinck and Johnson 2008).

4. Within coastal management and the fisheries sector, aquatic creatures are counted and categorized in different ways. In the fisheries sector, this can include categories based on export value or domestic consumption. These ideas of "fish" are tied to changing tastes and appetites (Probyn 2016).

5. These terms are not uniformly used across states and I am going by the way India's Department of Agriculture counts and classifies boats in its marine census. Midsize mechanized boats are clearly distinguished from trawlers in the annual surveys conducted by the state to measure the yearly catch. The state uses three primary categories: mechanized (this includes trawlers and other boats with onboard engines), motorized (smaller craft with outboard motors), and nonmechanized (craft operated by hand). It further classifies based on technologies and practices of fishing. The 2010 Marine Census of Maharashtra (2012, 4) notes: "There were 18,185 crafts owned by fisherfolk, of which 9,493 were mechanized, 1,346 motorized and 7,346 non-motorized. Dolnetters accounted for 47 percent of the mechanized crafts owned by fisherfolk, followed by gillnetters (32 percent) and trawlers 12 percent."

6. These are usually bought secondhand from small trucks. The Fisheries Department gives licenses based on horsepower. In recent times, a number of fishers are upgrading to higher power engines, which leads to tensions, as these more powerful motors lead to asymmetries in catch across boats (Devadasan and Boopendranath 2009).

7. Peke's (2013) study and Nair's (2021) book gives a detailed understanding of the tensions between Koli women, who traditionally worked as fish vendors, and migrant workers who come to work in the fish market. Nair's book also grounds this in relation to Mumbai's land and regional politics.

8. Raje and Deshmukh (1989) provide a detailed description of the process along with terminology.

9. This is described as fishing with "passive" gear. Although one can "actively" look for shoals of fish and capture them on determining their location, passive gear involves nets suspended in water, which rely on the movement of fish and on currents.

10. These decision-making processes are not uniform across small-scale fishing practices. For instance, recreational anglers think of individual fish and fish encounters, which is a completely different way of thinking-like-fish (Bear and Eden 2011) than bag-net fishers who might think in terms of aggregates.

11. This process was described to me—I have not actually seen it myself—but Sehara and Kharbari (1987) provide a detailed account of the mechanics of installing stakes. I did, however, see a stake when one was uninstalled and brought ashore.

12. The end of the conical bag net has another bit of net attached to it that has a finer mesh. This is called a sari and since the holes in it are much smaller, the small shrimp are caught in this part of the net. This design ensures they do not need to be sorted and can be directly transferred to tubs.

13. The buyers included retail and wholesale sellers.

14. Capture-fisheries refers to fishing in open waters (in a sea or a lake)—unlike farmed fish, which happens in a controlled setting.

15. The agents involved in the 2008 terror attacks on the city arrived via boat from the sea. After the event, the Indian coast guard launched a program to train fishers to identify potential threats. This program involves guards conducting security drills while posing as terrorists to see if the fishing boats will correctly identify the threat and report them.

16. *Annual Fish Production Report 2010–2011*, Fisheries Department, Government of Maharashtra, 2012.

17. *Annual Fish Production Report 2010–2011*, Fisheries Department, Government of Maharashtra, 2012.

18. Though, as Anita Maurstad (2002) writes, it is important to consider the ways in which fishers' knowledge is transferred to coastal management and to be mindful of the context in which these knowledge worlds live.

19. MoEF, Coastal Regulatory Zone Notification, Gazette of India, Extraordinary, Part-II, Section 3, Sub-section (ii), 1991.

20. See Narayanan 2014.

3. ANOMALOUS LANDSCAPES

1. Parts of this chapter have appeared as an article. See V. Chitra, "Remembering the River: Flood, Memory and Infrastructural Ecologies of Stormwater Drainage in Mumbai," *Urban Studies* 59 (9) (2022): 1855–71.

2. As historians of the monsoon and hydrological engineering in India assert, the monsoons were always rife with uncertainty. See Amrith 2018; and D'Souza 2016.

3. Rumors continued to spread in the days after the flood as people began hoarding medicines thought to cure flood-borne diseases. The MCGM and the Mumbai Police eventually launched a campaign to dispel these rumors (Fact Finding Committee Report 2006).

4. Randy Krum (2013), distinguishes between "data visualizations" and "infographics" thusly: while data visualizations focus on the numerical, infographics are combinations of several visualizations put together to create a story that can stand apart from the text. An infographic can be read as a standalone narrative.

5. The popularity of infographics is also a result of Internet-based weather information systems, which had, right before the flood, established a firm presence in the public domain. Data took on a different life after 2005 because they could circulate a lot more freely on the net.

6. IMD glossary of terms. Available at http://imd.gov.in/section/nhac/termglossary.pdf.

7. This information is available on the IMD's official website. It should be noted that the publication and broadcasting trends are rapidly changing. For example, the IMD has introduced a "nowcast" where information from its weather stations is available in real-time: https://mausam.imd.gov.in/imd_latest/contents/stationwise-nowcast-warning.php.

8. The inseparability of the weather and the physical city is manifestly evident through the research on postdisaster landscapes and politics. See Adams 2013; Klinenberg 2015; Murray 2009.

9. This is not unique to Mumbai, statistical analysis of the circulation of weather-related information shows that it does impact how people apprehend and react to local weather. For more information, see Kirilenko et al. 2015.

10. The problem of solid waste management and the history of Mumbai's dumping grounds are pertinent because of how infrastructures of garbage intersect with drainage, and I take this up in the fifth chapter.

11. Despite the fact that the drainage system in the south of the city is referred to as a "formal" system, a term that implies that it was engineered to work as a coordinated urban infrastructure, this network of pipes was not without its problems. The entire system was put in place over a long period as and when the need arose (Gupta 2007).

12. The Mithi is also divided into two parts that fall under different administrations. The top half from Vihar to the Bandra Kurla Complex is managed by the MCGM and the lower part falls under the purview of the MMRDA.

13. For example, see the Observer Research Foundation's report and recommendations titled, "Making the Mithi a River Again." Available online at https://www.indiawaterportal.org/sites/indiawaterportal.org/files/why_mumbai_must_reclaim_its_mithi_gautam_kirtane_orf_2011.pdf.

14. For example, Singapore and Mumbai receive, on average, almost the same amount of rainfall over a year (2340 mm and 2386 mm respectively). However, while the rain in Singapore is distributed over the span of ten months, Mumbai's rainfall occurs in intense bursts over a period of three months.

15. For example, extremely intense rainfall of 100 mm/hour might occur once in one hundred years in a specific place; that is, it has a 1 percent probability of occurring in any given year, whereas a 50mm/hr rainfall might occur once in ten years in the same location.

16. More information about this "movement" can be found on its Facebook page: https://www.facebook.com/rivermarchIND/.

4. UNFOLDING PLANS

1. Vinaja Punekar's (1959) study provides an account of the Koli community, including a map of the village at Versova.

2. As I note in the introduction, I had joined this school as visiting faculty and was able to participate in the architectural studio. However, I did not have any grading duties for this studio.

3. The Design Cell at KRVIA has documented this project and their successive work with fisher communities in Mumbai. A presentation on the studio exercise is available here: https://www.hmw.tu-berlin.de/fileadmin/i41_hmw/3_brazil_empowered_communities.pdf.

4. This does not mean there was no participation before the SRA. Participatory planning has a longer history—SRA helped promote and reinvent it as a practice. For example, community planning was linked to global flows of capital: it became more common for

World Bank funding programs to institute community participation as a part of its fund-ing aims, thus making it an integral aspect of many rehabilitation projects (Sanyal and Mukhija 2001).

5. A derogatory term used to discriminate against North Indian workers.

6. Original in Marathi, my translation.

5. "THIS IS NOT A DEAD ZONE"

1. Parts of this chapter were published as "Waste's Translations: Estuaries, Marine Life, and the Chemistry of Mumbai's Dumping Grounds," *American Ethnologist* 48 (4) (2021): 337–56.

2. The caste politics and urban dynamics within the Koli community (even within one village) are complex and shot through by the force of other relationships such as those formed through work and tenancies; they should not be taken as fixed or uniform across Mumbai's koliwadas.

3. Waste includes solid waste, sewage, and industrial effluents. While this article focuses primarily on solid waste, I use the term with the understanding that it refers to different kinds of complex mixes and I have tried to specify these where possible.

4. Kornberg (2019) writes about this in a study of transforming waste networks and technologies in Delhi.

5. For a detailed study, see Gidwani 2015.

6. Searle (2016) offers a detailed view of this transformation and the role of archi-tecture in it.

7. Sites that are not planned are called dumping grounds (I have also heard them described as open landfills), whereas the term *landfill* is typically used for sites engineered for storing waste. However, there seems to be some slippage between these terms.

8. There are many ways of generating energy, such as generating biogas and through incineration.

9. A large proportion of waste generated in India is organic waste, which has a high moisture content. Though this is a trend that could change as packaged food becomes cheaper and easily available, it is the reason why waste-to-energy programs are inefficient solutions and often fail (Luthra 2017).

10. This is not to obfuscate the fact that emanations from dumping grounds—liquid or gaseous—are hardly controlled and have disastrous consequences for waste workers and the residents of informal communities who live in close proximity (Salve et al. 2019; TISS 2015). It is only to explain the anthropocentrism that underlie these planned trans-formations, which often do not unfold as planned.

11. MLOM has moved to iNaturalist as a platform for archiving observations at the shore, though images also continue to be posted on its social media channels. Their archive is at https://www.marinelifeofmumbai.in/.

12. Even though the numbers are rising, Mumbai still produces far less waste per cap-ita than other cities such as Singapore or New York City: Mumbai produces approximately 06.kg per capita per day as against Singapore's 2kg per capita per day and NYC's 1.4 kg per capita per day. Moreover, its waste is mostly wet, organic garbage as opposed to the high percentage of plastic in cities like Singapore where on average, a person discards 13 plastic packages a day. Data available from NYC's Department of Sanitation's 2019 Annual Report and Singapore's National Environmental Agency. Down to Earth has a compre-hensive infographic on the distribution of waste in India (data from 2016): https://www.downtoearth.org.in/dte-infographics/clean_your_backyard-57790.html.

Bibliography

Abraham, C. M., and Sushila Abraham. 1991. "The Bhopal Case and the Development of Environmental Law in India." *International and Comparative Law Quarterly* 40 (2): 334–65.

Acheson, James M., and Roy Gardner. 2011. "Modeling Disaster: The Failure of the Management of the New England Groundfish Industry." *North American Journal of Fisheries Management* 31 (6): 1005–18. DOI: 10.1080/02755947.2011.635119.

Adams, Vincanne. 2013. *Markets of Sorrow, Labors of Faith: New Orleans in the Wake of Katrina.* Duke University Press.

Ahmann, Chloe. 2019. "Waste to Energy: Garbage Prospects and Subjunctive Politics in Late-Industrial Baltimore." *American Ethnologist* 46 (3): 328–42. DOI: 10.1111/amet.12792

Alaimo, Stacy. 2010. *Bodily Natures: Science, Environment, and the Material Self.* Bloomington: Indiana University Press.

Amrith, Sunil. 2018. *Unruly Waters: How Rains, Rivers, Coasts, and Seas Have Shaped Asia's History.* New York: Basic Books.

Anand, Nikhil. 2017. *Hydraulic City: Water and the Infrastructures of Citizenship in Mumbai.* Durham, NC: Duke University Press.

Anand, Nikhil, and Caroline Terens. 2019. "Explained: How Climate Change Could Impact Mumbai by 2050." *Indian Express*, December 13, Mumbai edition, sec. Explained. https://indianexpress.com/article/explained/the-rising-threat-to-mumbai-6160595/#:~:text=The%20research%2C%20carried%20by%20many,city%20has%20been%20built%20on.

Angelo, Hillary, and David Wachsmuth. 2015. "Urbanizing Urban Political Ecology: A Critique of Methodological Cityism." *International Journal of Urban and Regional Research* 39 (1): 16–27. DOI: 10.1111/1468-2427.12105.

Anjaria, Jonathan Shapiro. 2009. "Guardians of the Bourgeois City: Citizenship, Public Space, and Middle-Class Activism in Mumbai 1." *City and Community* 8 (4): 391–406. 10.1111/j.1540-6040.2009.0129

Annual Fish Production Report 2010–2011. 2012. Fisheries Department, Government of Maharashtra.

Appadurai, Arjun. 2001. "Deep Democracy: Urban Governmentality and the Horizon of Politics." *Environment and Urbanization* 13 (2): 23–43. DOI: 0.1177/095624780101300203.

Arunachalam, B. 2005. "Drainage Problems of Brihan Mumbai." *Economic and Political Weekly* 40 (36): 3909–11. https://www.jstor.org/stable/4417102.

Attarwala, F. N. 1994. "Solid Waste Management in Greater Bombay: Final Report." Mumbai: National Environmental Engineering Institute.

Bakker, Karen. 2010. "The Limits of 'Neoliberal Natures': Debating Green Neoliberalism." *Progress in Human Geography* 34 (6): 715–35. DOI: 10.1177/0309132510376849.

Balakrishnan, Sai. 2019. *Shareholder Cities: Land Transformations along Urban Corridors in India.* Philadelphia, PA: University of Pennsylvania Press.

Ballestero, Andrea. 2019. *A Future History of Water*. Durham, NC: Duke University Press.

Barry, Lynda. 2019. *Making Comics*. Illustrated ed. Montréal, Canada: Drawn and Quarterly.

Bavinck, Maarten, and Derek Johnson. 2008. "Handling the Legacy of the Blue Revolution in India: Social Justice and Small-Scale Fisheries in a Negative Growth Scenario." In *American Fisheries Society Symposium*, 49:585. American Fisheries Society. Bethesda, MD.

Bavington, Dean. 2011. *Managed Annihilation: An Unnatural History of the Newfoundland Cod Collapse*. Vancouver: University of British Columbia Press.

Baviskar, Amita. 2020. "Cows, Cars, and Cycle-Rickshaws: Bourgeois Environmentalists and the Battle for Delhi's Streets." In *Elite and Everyman*, edited by Amita Bhide and Raka Ray, 391–418. New Delhi: Routledge India.

Bear, Christopher, and Sally Eden. 2011. "Thinking Like a Fish? Engaging with Nonhuman Difference through Recreational Angling." *Environment and Planning D: Society and Space* 29 (2): 336–52. DOI: 10.1068/d1810.

Beaulieu, Anne. 2014. "If Visual STS Is the Answer, What Is the Question?" In *Visualization in the Age of Computerization*, edited by Annemaria Carusi, Aud Sissel Hoel, Timothy Webmoor, and Steve Woolgar, 237–42. New York: Routledge.

Benjamin, Solomon. 2010. "Manufacturing Neoliberalism: Lifestyling Indian Urbanity." In *Accumulation by Dispossession: Transformative Cities in the New Global Order*, edited by Swapna Banerjee-Guha, 92–124. New Delhi: SAGE Publications.

Bennett, Jane. 2010. *Vibrant Matter: A Political Ecology of Things*. Durham, NC: Duke University Press.

Bhagat, R. B., Mohua Guha, and Aparajita Chattopadhyay. 2006. "Mumbai after 26/7 Deluge: Issues and Concerns in Urban Planning." *Population and Environment* 27 (4): 337–49. DOI: 10.1007/s11111-006-0028-z.

Bhalla, R. S. 2007. "Do Bio-Shields Affect Tsunami Inundation?" *Current Science* 93 (26): 831–33. https://www.jstor.org/stable/24099129.

Bhattacharyya, Debjani. 2018. *Empire and Ecology in the Bengal Delta: The Making of Calcutta*. Cambridge: Cambridge University Press.

Bhide, Amita. 2023. "Structural Violence in Much More Than Neoliberal Times: The Case of Slum Redevelopment in Mumbai." *City*, 1–18. DOI: 10.1080/13604813.2023.2223880.

Bhutia, Lhendup. 2020. "Coral Relief." *Open the Magazine*, November 20.

Björkman, Lisa. 2015. *Pipe Politics, Contested Waters: Embedded Infrastructures of Millennial Mumbai*. Durham, NC: Duke University Press.

Björkman, Lisa, and Andrew Harris. 2018. "Engineering Cities: Mediating Materialities, Infrastructural Imaginaries, and Shifting Regimes of Urban Expertise." *International Journal of Urban and Regional Research* 42 (2): 244–62. DOI: 10.1111/1468-2427.12528.

Bowen, Robert E., and Cory Riley. 2003. "Socio-Economic Indicators and Integrated Coastal Management." *Ocean and Coastal Management* 46 (3–4): 299–312. DOI: 10.1016/S0964-5691(03)00008-5.

Bowonder, B. 1986. "Environmental Management Problems in India." *Environmental Management* 10 (5): 599–609. DOI: 10.1007/BF01866764.

Boyer, Dominic. 2014. "Energopower: An Introduction." *Anthropological Quarterly* 87 (2): 309–33. DOI: https://www.jstor.org/stable/43652700.

Bremner, Lindsay. 2020. "Planning the 2015 Chennai Floods." *Environment and Planning E: Nature and Space* 3 (3): 732–60. DOI: 10.1177/2514848619880130.

"BRIMSTOWAD: Brihanmumbai Storm Water Drainage Report." 1993. Mumbai: Municipal Corporation of Greater Mumbai.

Burke, Lauretta, Yumiko Kura, Ken Kassem, Carmen Revenga, Mark Spalding, and Don McAllister. 2001. *Pilot Analysis of Global Coastal Ecosystems.* Washington, DC: World Resources Institute.

Burri, Regula Valérie, and Joseph Dumit. 2008. "Social Studies of Scientific Imaging and Visualization." In *The Handbook of Science and Technology Studies*, 3rd ed., edited by Edward Hackett, Olga Amsterdamska, Michael E. Lynch, and Judy Wajcman, 297–317. Cambridge, MA: MIT Press.

Butt, Waqas H. 2020. "Waste Intimacies: Caste and the Unevenness of Life in Urban Pakistan." *American Ethnologist* 47 (3): 234–48.

Campling, Liam, and Alejandro Colás. 2018. "Capitalism and the Sea: Sovereignty, Territory and Appropriation in the Global Ocean." *Environment and Planning D: Society and Space* 36 (4): 776–94. DOI: 10.1177/0263775817737.

Carruth, Allison, and Robert P. Marzec. 2014. "Environmental Visualization in the Anthropocene: Technologies, Aesthetics, Ethics." *Public Culture* 26 (2/73): 205–11. DOI: 10.1215/08992363-2392030.

Center for Science and Environment. 2017. "Mumbai," August 31. https://www.cseindia.org/mumbai-1751.

Chakrabarty, Dipesh. 2009. "The Climate of History: Four Theses." *Critical Inquiry* 35 (2): 197–222. DOI: 10.1111/amet.12960.

Chao, Sophie. 2022. *In the Shadow of the Palms: More-Than-Human Becomings in West Papua.* Durham, NC: Duke University Press.

Chatfield, Akemi Takeoka, Hans J. Jochen Scholl, and Uuf Brajawidagda. 2013. "Tsunami Early Warnings via Twitter in Government: Net-Savvy Citizens' Co-Production of Time-Critical Public Information Services." *Government Information Quarterly* 30 (4): 377–86. DOI: 10.1016/j.giq.2013.05.021.

Chatterjee, Badri. 2018. "105% Rise in Waste Generated in Mumbai from 1999 to 2016: Study." *Hindustan Times*, August 5, 2018.

Chatterjee, Badri. 2019a. "High Court Stays Expansion of Kanjurmarg Dumping Ground in Mumbai." *Hindustan Times*, September 21, 2019. https://www.hindustantimes.com/mumbai-news/high-court-stays-expansion-of-kanjurmarg-dumping-ground-in-mumbai/story-HSHxrKspjzWqy1WGtzSerL.html.

Chatterjee, Badri. 2019b. "95% of Mumbai's Mangroves Could Perish from Rising Sea Levels: State." *Hindustan Times*, November 16, 2019.

Chatterjee, Partha. 2010. *Empire and Nation.* New York: Columbia University Press.

Chatterjee, Syantani. 2019. "The Labors of Failure: Labor, Toxicity, and Belonging in Mumbai." *International Labor and Working-Class History* 95: 49–75. DOI: 10.1017/S0147547919000073.

Chhabria, Sheetal. 2018. "The Aboriginal Alibi: Governing Dispossession in Colonial Bombay." *Comparative Studies in Society and History* 60 (4): 1096–1126. DOI: 10.1017/S0010417518000397.

Chion, Michel. 2019. *Audio-Vision: Sound on Screen.* New York: Columbia University Press.

Chokhandre, Praveen, Shrikant Singh, and Gyan Chandra Kashyap. 2017. "Prevalence, Predictors and Economic Burden of Morbidities among Waste-Pickers of Mumbai, India: A Cross-Sectional Study." *Journal of Occupational Medicine and Toxicology* 12 (1): 30. DOI: 10.1186/s12995-017-0176-3.

Chu, Julie Y. 2014. "When Infrastructures Attack: The Workings of Disrepair in China." *American Ethnologist* 41 (2): 351–67. DOI: 10.1111/amet.12080.

"Coastal Road Mumbai: Report by the Joint Technical Committee." 2011. Government of Maharashtra, December.

Coelho, Karen, Lalitha Kamath, and M. Vijayabaskar. 2013. "Opening up or Ushering in? Citizen Participation as Mandate and Practice in Urban Governance." In *Participolis: Consent and Contention in Neoliberal Urban India*, 1st ed., edited by Karen Coelho, Lalitha Kamath, and M. Vijayabaskar, 3–33. New Delhi: Routledge India.

Coelho, Karen, and Nithya V. Raman. 2010. "Salvaging and Scapegoating: Slum Evictions on Chennai's Waterways." *Economic and Political Weekly* 45 (21): 19–23. https://www.jstor.org/stable/27807045.

Concerned Citizens' Commission. 2006. "Mumbai Marooned: An Enquiry into Mumbai Floods 2005." Mumbai, India: Conservation Action Trust (CAT). https://cat.org.in/wp-content/uploads/2017/03/Mumbai-Marooned-An-Enquiry-into-Mumbais-Floods-2005.pdf.

Corburn, Jason. 2003. "Bringing Local Knowledge into Environmental Decision Making: Improving Urban Planning for Communities at Risk." *Journal of Planning Education and Research* 22 (4): 420–33. DOI: 10.1177/0739456X03022004008.

Costa, Joao Pinto da, Patricia SM Santos, Armando C. Duarte, and Teresa Rocha-Santos. 2016. "(Nano) Plastics in the Environment–Sources, Fates, and Effects." *Science of the Total Environment* 566: 15–26. DOI: 10.1016/j.scitotenv.2016.05.041.

Cousins, Joshua J. 2017. "Volume Control: Stormwater and the Politics of Urban Metabolism." *Geoforum* 85: 368–80. 10.1016/j.geoforum.2016.09.020.

CPCB (Central Pollution Control Board). 2016. "Solid Waste Generation in 46 Metro-Cities." India.

Crampton, Jeremy W. 2009. "Cartography: Maps 2.0." *Progress in Human Geography* 33 (1): 91–100. DOI: 10.1177/0309132508094074.

Cruikshank, Julie. 2007. *Do Glaciers Listen?: Local Knowledge, Colonial Encounters, and Social Imagination*. Vancouver: University of British Columbia Press.

"CRZ Notification 1991–2006: Saga of Betrayal-Andaman and Nicobar Islands." 2007. Bengaluru, Karnataka: Equitable Tourism Options (EQUATIONS).

Daston, Lorraine. 2019. "The Coup d'Oeil: On a Mode of Understanding." *Critical Inquiry* 45 (2): 307–31. DOI: 10.1086/700990.

Daston, Lorraine, and Peter Galison. 2010. *Objectivity*. Princeton, NJ: Princeton University Press.

Degnbol, Poul, and Bonnie J. McCay. 2007. "Unintended and Perverse Consequences of Ignoring Linkages in Fisheries Systems." *ICES Journal of Marine Science* 64 (4): 793–97. DOI: 10.1093/icesjms/fsm040.

Devadasan, K., and M. R. Boopendranath. 2009. "Fishing Craft and Gear for Small Pelagics." Paper presented at the *National Conference on Marine Fisheries and Fisheries Harbour Infrastructure*, February 7–8, 2009, Fishery Survey of India, Mumbai, 7–8.

Dewan, Camelia. 2021. *Misreading the Bengal Delta: Climate Change, Development, and Livelihoods in Coastal Bangladesh*. Seattle: University of Washington Press.

Dharia, Namita Vijay. 2022. *The Industrial Ephemeral: Labor and Love in Indian Architecture and Construction*. Berkeley: University of California Press.

Doctor, Vikram. 2014. "Jellyfish Blooms and Ganapati Immersions." *Economic Times*, September. https://economictimes.indiatimes.com/blogs/onmyplate/jellyfish-blooms-and-ganapati-immersions/.

Doherty, Jacob. 2019. "Filthy Flourishing: Parasites, Animal Infrastructure, and the Waste Frontier in Kampala." *Current Anthropology* 60 (S20): S321–32. DOI: 10.1086/702868.

Doron, Assa, and Robin Jeffrey. 2018. *Waste of a Nation: Garbage and Growth in India*. Cambridge, MA: Harvard University Press.

Doshi, Sapana. 2013a. "The Politics of the Evicted: Redevelopment, Subjectivity, and Difference in Mumbai's Slum Frontier." *Antipode* 45 (4): 844–65. DOI: 10.1111/j.1467–8330.2012.01023.x.

Doshi, Sapana. 2013b. "Resettlement Ecologies: Environmental Subjectivity and Graduated Citizenship in Mumbai." *Ecologies of Urbanism in India: Metropolitan Civility and Sustainability*, 225–48. DOI: 10.5790/hongkong/9789888139767.003.0009.

Dossal, Mariam. 2005. "A Master Plan for the City: Looking at the Past." *Economic and Political Weekly* 40 (36): 3897–3900. https://www.jstor.org/stable/4417098.

Dossal, Mariam. 2010. *Theatre of Conflict, City of Hope: Mumbai 1660 to Present Times*. New Delhi: Oxford University Press.

Douglas, Mary. 2002. *Risk and Blame*. London: Routledge.

Douglas-Jones, Rachel. 2021. "Drawing as Analysis: Thinking in Images and Writing in Words." In *Experimenting with Ethnography: A Companion to Analysis*, edited by Andrea Ballestero and Brit Ross Winthereik, 94–105. Durham, NC: Duke University Press.

D'Souza, Rohan. 2016. *Drowned and Dammed: Colonial Capitalism and Flood Control in Eastern India*. New Delhi: Oxford University Press.

Dupont, Véronique, Marie-Caroline Saglio-Yatzimirsky, and Lenise Lima Fernandes. 2013. "Public Policies and the 'Treatment' of Slums." In *Megacity Slums: Social Exclusion, Space, and Urban Policies in Brazil and India*, edited by Marie-Caroline Saglio-Yatzimirsky and Frédéric Landy, 165–212. London: Imperial College Press.

Dutta, Abhishek, and Wanida Jinsart. 2020. "Waste Generation and Management Status in the Fast-Expanding Indian Cities: A Review." *Journal of the Air and Waste Management Association* 70 (5): 491–503. DOI: 10.1080/10962247.2020.1738285.

Dwarakish, G. S., S. A. Vinay, S. M. Dinakar, J. B. Pai, K. Mahaganesh, and U. Natesan. 2008. "Integrated Coastal Zone Management Plan for Udupi Coast Using Remote Sensing, Geographical Information System and Global Position System." *Journal of Applied Remote Sensing* 2 (1): 023515-17. DOI:10.1117/1.2919101.

Dwivedi, Sharada, and Rahul Mehrotra. 1995. *Bombay: The Cities Within*. Mumbai: India Book House.

Edney, Matthew H. 1997. *Mapping an Empire: The Geographical Construction of British India, 1765–1843*. Chicago: University of Chicago Press.

Edwardes, Stephen Meredyth. 1912. *By-Ways of Bombay*. Bombay: DB Taraporevala Sons.

Elden, Stuart. 2013. "Secure the Volume: Vertical Geopolitics and the Depth of Power." *Political Geography* 34: 35–51.

Elias, Ann. 2019. *Coral Empire: Underwater Oceans, Colonial Tropics, Visual Modernity*. Durham, NC: Duke University Press.

FAO (Food and Agricultural Organization). 2022. "The State of World Fisheries and Aquaculture 2022. Towards Blue Transformation." Rome: FAO. DOI: 10.4060/cc0461en.

Fentiman, Alicia. 1996. "The Anthropology of Oil: The Impact of the Oil Industry on a Fishing Community in the Niger Delta." *Social Justice* 23 4 (66): 87–99. https://www.jstor.org/stable/29766976.

Finkelstein, Maura. 2018. "Ghosts in the Gallery: The Vitality of Anachronism in a Mumbai Chawl." *Anthropological Quarterly* 91 (3): 937–68. ISSN 0003-5491.

"Fish Production Report 2010–2011." 2011. Mumbai: Department of Fisheries, Government of Maharashtra.

"Fisheries Profile of India." 2018. Department of Fisheries. Government of India.

Gabbott, Sarah, Sarah Key, Catherine Russell, Yasmin Yohan, and Jan Zalasiewicz. 2020. "The Geography and Geology of Plastics: Their Environmental Distribution and Fate." In *Plastic Waste and Recycling: Environmental Impact, Societal Issues, Prevention, and Solutions*, edited by Trevor Letch, 33–63. Elsevier. DOI: 10.1016/B978-0-12-817880-5.00003–7.

Gandolfo, Daniella. 2013. "Formless: A Day at Lima's Office of Formalization." *Cultural Anthropology* 28 (2): 278–98. DOI: 10.1111/cuan.12004.

Gandy, Matthew. 2008. "Landscapes of Disaster: Water, Modernity, and Urban Fragmentation in Mumbai." *Environment and Planning A* 40 (1): 108–30. DOI: 10.1068/a3994.

García-Quijano, Carlos G. 2007. "Fishers' Knowledge of Marine Species Assemblages: Bridging Between Scientific and Local Ecological Knowledge in Southeastern Puerto Rico." *American Anthropologist* 109 (3): 529–36. DOI: 10.1525/aa.2007.109.3.529.

Ghertner, D. Asher. 2017. "When Is the State? Topology, Temporality, and the Navigation of Everyday State Space in Delhi." *Annals of the American Association of Geographers* 107 (3): 731–50. DOI: 10.1080/24694452.2016.1261680.

Ghosh, Amitav. 2016. *The Great Derangement: Climate Change and the Unthinkable.* Chicago: University of Chicago Press.

Gidwani, Vinay. 2015. "The Work of Waste: Inside India's Infra-Economy." *Transactions of the Institute of British Geographers* 40 (4): 575–95. DOI: 10.1111/tran.12094.

Gidwani, Vinay, and Julia Eleanor Corwin. 2017. "Governance of Waste." *Economic and Political Weekly* 52 (31): 44–54. https://www.jstor.org/stable/26695935.

Giraud, Eva Haifa. 2019. *What Comes after Entanglement?: Activism, Anthropocentrism, and an Ethics of Exclusion.* Durham, NC: Duke University Press.

Goh, Kian. 2019. "Urban Waterscapes: The Hydro-Politics of Flooding in a Sinking City." *International Journal of Urban and Regional Research* 43 (2): 250–72.

Gordillo, Gastón. 2006. "The Crucible of Citizenship: ID-Paper Fetishism in the Argentinean Chaco." *American Ethnologist* 33 (2): 162–76. DOI: 10.1525/ae.2006.33.2.162.

Govindrajan, Radhika. 2019. *Animal Intimacies: Interspecies Relatedness in India's Central Himalayas.* Chicago: University of Chicago Press.

Green, Lesley. 2020. *Rock| Water| Life.* Durham, NC: Duke University Press.

Groves, Christopher. 2017. "Emptying the Future: On the Environmental Politics of Anticipation." *Futures* 92: 29–38. DOI: 10.1016/j.futures.2016.06.003.

Guhathakurta, P., O. P. Sreejith, and P. A. Menon. 2011. "Impact of Climate Change on Extreme Rainfall Events and Flood Risk in India." *Journal of Earth System Science* 120 (3): 359. DOI: 0.1007/s12040-011-0082-5.

Gupta, Kapil. 2007. "Urban Flood Resilience Planning and Management and Lessons for the Future: A Case Study of Mumbai, India." *Urban Water Journal* 4 (3): 183–94. DOI: 10.1080/15730620701464141.

Gupte, A. K. 2009. *City Survey Manual with Explanatory Notes [English]*. Mumbai: Mukund Prakashan.

Guyer, Jane I. 2016. *Legacies, Logics, Logistics: Essays in the Anthropology of the Platform Economy*. Chicago: University of Chicago Press.

Guyer, Jane I., Naveeda Khan, Juan Obarrio, Caroline Bledsoe, Julie Chu, Souleymane Bachir Diagne, Keith Hart, Paul Kockelman, Jean Lave, and Caroline McLoughlin. 2010. *Introduction: Number as Inventive Frontier*. London: Sage.

Halpern, Orit. 2015. *Beautiful Data: A History of Vision and Reason since 1945*. Durham, NC: Duke University Press.

Hamdy, Sherine, and Coleman Nye. 2017. *Lissa: A Story about Medical Promise, Friendship, and Revolution*. Illustrated ed. North York, Canada: University of Toronto Press.

"Handbook of Fisheries Statistics 2014." 2014. Ministry of Fisheries, Animal Husbandry, and Dairying, Government of India.

"Handbook of Fisheries Statistics 2020." 2020. Ministry of Fisheries, Animal Husbandry, and Dairying, Government of India.

Hansen, Thomas Blom. 2001. *Wages of Violence: Naming and Identity in Postcolonial Bombay*. Princeton, NJ: Princeton University Press.

Haraway, Donna. 1988. "Situated Knowledges: The Science Question in Feminism and the Privilege of Partial Perspective." *Feminist Studies* 14 (3): 575–99. DOI: 10.2307/3178066.

Haraway, Donna. 1992. "The Promises of Monsters: A Regenerative Politics for Inappropriate/d Others." In Cultural Studies, edited by Lawrence Grossberg, 295–337. New York and London: Routledge.

Harley, John Brian. 2002. *The New Nature of Maps: Essays in the History of Cartography*. 2002. Baltimore: Johns Hopkins University Press.

Harms, Erik. 2020. "The Case of the Missing Maps: Cartographic Action in Ho Chi Minh City." *Critical Asian Studies* 52 (3): 332–63. DOI: 10.1080/14672715.2020.1792784.

Harris, Andrew J. L., and Massimo Lanfranco. 2017. "Cloudburst, Weather Bomb or Water Bomb? A Review of Terminology for Extreme Rain Events and the Media Effect." *Weather* 72 (6): 155–63. DOI: 10.1002/wea.2923.

Helmreich, Stefan. 2011. "From Spaceship Earth to Google Ocean: Planetary Icons, Indexes, and Infrastructures." *Social Research* 78 (4): 1211–42. DOI:10.1353/sor.2011.0042.

Helmreich, Stefan. 2014. "Waves: An Anthropology of Scientific Things." *HAU: Journal of Ethnographic Theory* 4 (3): 265–84. DOI: 10.14318/hau4.3.016.

Henderson, Kathryn. 1998. *On Line and on Paper: Visual Representations, Visual Culture, and Computer Graphics in Design Engineering*. Cambridge, MA: MIT Press.

Hetherington, Kregg. 2011. *Guerrilla Auditors*. Durham, NC: Duke University Press.

Hird, Myra J. 2013. "Waste, Landfills, and an Environmental Ethic of Vulnerability." *Ethics and the Environment* 18 (1): 105–24. DOI: 10.2979/ethicsenviro.18.1.105.

Hoag, Colin, Filippo Bertoni, and Nils Bubandt. 2018. "Wasteland Ecologies: Undomestication and Multispecies Gains on an Anthropocene Dumping Ground." *Journal of Ethnobiology* 38 (1): 88–105. DOI: 10.2993/0278-0771-38.1.088.

Hoeppe, Götz. 2007. *Conversations on the Beach: Fishermans's Knowledge, Metaphor, and Environmental Change In South India.* Vol. 2. New York: Berghahn Books.

Houghton, John Theodore, YDJG Ding, David J. Griggs, Maria Noguer, Paul J. van der Linden, Xiaosu Dai, Kathy Maskell, and C. A. Johnson. 2001. *Climate Change 2001: The Scientific Basis.* Oxford: Press Syndicate University of Cambridge.

Houser, Heather. 2014. "The Aesthetics of Environmental Visualizations: More than Information Ecstasy?" *Public Culture* 26 (2): 319–37. DOI:10.1215/08992363-2392084.

Howe, Cymene. 2014. "Anthropocenic Ecoauthority: The Winds of Oaxaca." *Anthropological Quarterly* 87 (2): 381–404. https://www.jstor.org/stable/43652703.

Howe, Cymene. 2019. "Sensing Asymmetries in Other-than-Human Forms." *Science, Technology, and Human Values* 44 (5): 900–910. DOI: 10.1177/0162243919852675.

Hull, Matthew S. 2012. *Government of Paper: The Materiality of Bureaucracy in Urban Pakistan.* Berkeley: University of California Press.

ICSF (International Collective in Support of Fishworkers). 2014. "Mumbai Fish Markets: A Mapping Exercise." Chennai, India. Available here: https://www.icsf.net/wp-content/uploads/2016/08/930.ICSF009.pdf.

Indian Express. 2012. "BMC Orders Probe into Carbon Credit Estimate Error," September 25. https://indianexpress.com/article/cities/mumbai/bmc-orders-probe-into-carbon-credit-estimate-error/.

Ingold, Tim. 2010. "Bringing Things to Life: Creative Entanglements in a World of Materials." *ESRC National Centre for Research Methods.*

Isaac, T. M., and Richard W. Franke. 2002. *Local Democracy and Development: The Kerala People's Campaign for Decentralized Planning.* Lanham, MD: Rowman & Littlefield.

Jadhav, Adam. 2018. "Undefining Small-Scale Fisheries in India: Challenging Simplifications and Highlighting Diversity and Value." In *Social Wellbeing and the Values of Small-Scale Fisheries,* 1st ed., edited by Derek S. Johnson, Tim G. Acott, Natasha Stacey, Julie Urquhart, 147–73. Amsterdam: Springer, MARE Publication Series 17. DOI: 10.1007/978-3-319-60750-4_7.

Jain, Kajri. 2007. *Gods in the Bazaar.* Durham, NC: Duke University Press.

James, C. C. 1906. *Drainage Problems of the East: Being a Revised and Enlarged Edition of "Oriental Drainage."* Mumbai, India: "Times of India" Office.

Jasanoff, Sheila. 1993. "India at the Crossroads in Global Environmental Policy." *Global Environmental Change* 3 (1): 32–52. DOI: 10.1016/0959-3780(93)90013-B.

Jasanoff, Sheila. 2004. "Heaven and Earth: The Politics of Environmental Images." In *Earthly Politics: Local and Global in Environmental Governance,* edited by Sheila Jasanoff and Marybeth Long Martello, 31–52. Cambridge, MA: MIT Press.

Jay, Martin. 2014. "Conclusion: A Parting Glance: Empire and Visuality." In *Empires of Vision: A Reader,* edited by Martin Jay and Sumathi Ramaswamy, 609–20. Durham, NC: Duke University Press.

Jayasiri, H. B., C. S. Purushothaman, and A. Vennila. 2013. "Quantitative Analysis of Plastic Debris on Recreational Beaches in Mumbai, India." *Marine Pollution Bulletin* 77 (1): 107–12. DOI: 10.1016/j.marpolbul.2013.10.024.

Jenamani, Rajendra Kumar, S. C. Bhan, and S. R. Kalsi. 2006. "Observational/Forecasting Aspects of the Meteorological Event That Caused a Record Highest Rainfall in Mumbai." *Current Science,* 1344–62. http://www.jstor.org/stable/24091984.

Jose, George. "Hawa khaana in Vasai Virar: Urban Housing and the Commodification of Airspace in Mumbai's Periphery." *City* 21 (5): 632–40.

Kain, Roger J. P., and Elizabeth Baigent. 1992. *The Cadastral Map in the Service of the State: A History of Property Mapping*. Chicago: University of Chicago Press.

Kamath, Lalitha, and Gopal Dubey. 2020. "Commoning the Established Order of Property: Reclaiming Fishing Commons in Mumbai." *Urbanisation* 5 (2): 85–101. DOI: 10.1177/245574712097298.

Kamath, Lalitha, and Marina Joseph. 2015. "How a Participatory Process Can Matter in Planning the City." *Economic and Political Weekly*: 54–61. https://www.jstor.org/stable/24482461.

Kaur, Raminder. 2003. *Performative Politics and the Cultures of Hinduism: Public Uses of Religion in Western India*. Hyderabad: Orient Blackswan.

Kesavan, P. C., and M. S. Swaminathan. 2007. "The 26 December 2004 Tsunami Recalled: Science and Technology for Enhancing Resilience of the Andaman and Nicobar Islands Communities." *Current Science* 92 (6): 743–47.

Khanolkar, Prasad. 2021. "Mehmoodbhai Toilet Operator." In *Bombay Brokers*, edited by Lisa Björkman, 87–94. Durham, NC: Duke University Press.

Kirilenko, Andrei P., Tatiana Molodtsova, and Svetlana O. Stepchenkova. 2015. "People as Sensors: Mass Media and Local Temperature Influence Climate Change Discussion on Twitter." *Global Environmental Change* 30: 92–100. DOI: 10.1016/j.gloenvcha.2014.11.003.

Klean Environmental Consultants. 2006. "Report on Pollution Study of Mithi River Basin." Mumbai: Maharashtra Pollution Control Board.

Klinenberg, Eric. 2015. *Heat Wave: A Social Autopsy of Disaster in Chicago*. Chicago: University of Chicago Press.

Kockelman, Paul, and Anya Bernstein. 2012. "Semiotic Technologies, Temporal Reckoning, and the Portability of Meaning. Or: Modern Modes of Temporality–Just How Abstract Are They?" *Anthropological Theory* 12 (3): 320–48. DOI: 10.1177/1463499612463308.

Kornberg, Dana. 2019. "Garbage as Fuel: Pursuing Incineration to Counter Stigma in Postcolonial Urban India." *Local Environment* 24 (1): 1–17. DOI: 10.1080/13549839.2018.1545752.

Krum, Randy. 2013. *Cool Infographics: Effective Communication with Data Visualization and Design*. Hoboken, NJ: John Wiley & Sons.

Kumar, Mukul, K. Saravanan, and Nityanand Jayaraman. 2014. "Mapping the Coastal Commons: Fisherfolk and the Politics of Coastal Urbanisation in Chennai." *Economic and Political Weekly*: 46–53. https://www.jstor.org/stable/24481080.

Kurien, John, and TR Thankappan Achari. 1990. "Overfishing Along Kerala Coast: Causes and Consequences." *Economic and Political Weekly* 5 (35/36): 2011–18. https://www.jstor.org/stable/4396716.

Lakoff, Andrew. 2017. *Unprepared: Global Health in a Time of Emergency*. Berkeley: University of California Press.

Latour, Bruno. 1986. "Visualization and Cognition." *Knowledge and Society* 6 (6): 1–40. ISBN: 0-89232-664-6.

Latour, Bruno. 1999. *Pandora's Hope: Essays on the Reality of Science Studies*. Cambridge, MA: Harvard University Press.

Latour, Bruno, and Steve Woolgar. 2013. *Laboratory Life: The Construction of Scientific Facts*. Princeton, NJ: Princeton University Press.

Lee, Hyung K. 2005. "Mapping the Law of Legalizing Maps: The Implications of the Emerging Rule of Map Evidence in International Law." *Pacific Rim Law and Policy Journal Association* 14: 159.

Lekan, Thomas M. 2014. "Fractal Earth: Visualizing the Global Environment in the Anthropocene." *Environmental Humanities* 5 (1): 171–201. DOI: 10.1215/22011919-3615469.

Ley, Lukas. 2018. "Discipline and Drain: River Normalization and Semarang's Fight against Tidal Flooding." *Indonesia* 105: 53–75. DOI: 10.1353/ind.2018.0002.

Li, Tania Murray. 2000. "Articulating Indigenous Identity in Indonesia: Resource Politics and the Tribal Slot." *Comparative Studies in Society and History* 42 (1): 149–79. DOI: 10.1017/S0010417500002632.

Liboiron, Max. 2015. "Disaster Data, Data Activism: Grassroots Responses to Representing Superstorm Sandy." In *Extreme Weather and Global Media*, edited by Julia Leyda and Diane Negra, 144–62. New York: Routledge.

Luthra, Aman. 2017. "Waste-to-Energy and Recycling: Competing Systems of Waste Management in Urban India." *Economic and Political Weekly* 52 (13): 51. DOI: https://www.jstor.org/stable/26695669.

Mansfield, Becky. 2004. "Neoliberalism in the Oceans: 'Rationalization,' Property Rights, and the Commons Question." *Geoforum* 35 (3): 313–26. DOI: 10.1016/j.geoforum.2003.05.002.

"Marine Census 2010 Maharashtra." 2012. Census Report. Kochi, New Delhi: Central Marine Fisheries Research Institute and Department of Animal Husbandry, Dairying, and Fisheries, Government of India.

Marino, Elizabeth. 2015. *Fierce Climate, Sacred Ground: An Ethnography of Climate Change in Shishmaref, Alaska*. Fairbanks: University of Alaska Press.

Martin, Kevin St, Bonnie J. McCay, Grant D. Murray, Teresa R. Johnson, and Bryan Oles. 2007. "Communities, Knowledge and Fisheries of the Future." *International Journal of Global Environmental Issues* 7 (2–3): 221–39. DOI: 10.1504/ijgenvi.2007.013575.

Martinussen, Einar Sneve, Jørn Knutsen, and Timo Arnall. 2014. "Satellite Lamps." *Kairos* 19 (1). Available at https://kairos.technorhetoric.net/19.1/inventio/martinussen-et-al/index.html.

Marzec, Robert P. 2015. *Militarizing the Environment: Climate Change and the Security State*. Minneapolis: University of Minnesota Press.

Mascarenhas, Antonio, and Seelam Jayakumar. 2008. "An Environmental Perspective of the Post-Tsunami Scenario along the Coast of Tamil Nadu, India: Role of Sand Dunes and Forests." *Journal of Environmental Management* 89 (1): 24–34. DOI: 10.1016/j.jenvman.2007.01.053.

Mathur, Anuradha, and Dilip Da Cunha. 2009. *SOAK: Mumbai in an Estuary*. New Delhi: Rupa & Company.

Mathur, Nayanika. 2016. *Paper Tiger*. Cambridge: Cambridge University Press.

Maurstad, Anita. 2002. "Fishing in Murky Waters—Ethics and Politics of Research on Fisher Knowledge." *Marine Policy* 26 (3): 159–66.

Mazzarella, William. 2010. "Beautiful Balloon: The Digital Divide and the Charisma of New Media in India." *American Ethnologist* 37 (4): 783–804. DOI: 10.1111/j.1548-1425.2010.01285.x.

McCay, Bonnie J. 2008. "The Littoral and the Liminal: Challenges to the Management of the Coastal and Marine Commons." Maritime Studies (MAST) 7 (1): 7–30.

McCloud, Scott. 1994. *Understanding Comics: The Invisible Art*. New York: William Morrow Paperbacks.

McFarlane, Colin. 2008. "Sanitation in Mumbai's Informal Settlements: State, 'Slum,' and Infrastructure." *Environment and Planning A* 40 (1): 88–107. DOI: 10.1068/a3922.

McFarlane, Colin. 2012. "The Entrepreneurial Slum: Civil Society, Mobility, and the Co-Production of Urban Development." *Urban Studies* 49 (13): 2795–2816. DOI: 10.1177/00420980124524.

Menon, Manju, Meenakshi Kapoor, P. Venkatraam, Kanchi Kohli, and S. Kumar. 2015. "CZMAs and Coastal Environments: Two Decades of Regulating Land Use Change on India's Coastline." Center for Policy Research-Namati, New Delhi.

Menon, Manju, and Kanchi Kohli. 2008. "Re-Engineering the Legal and Policy Regimes on Environment." *Economic and Political Weekly* 43 (23): 14–17. https://www.jstor.org/stable/40277536.

Menon, Manju, and Aarthi Sridhar. 2007. "An Appraisal of Coastal Regulation Law in Tsunami-Affected Mainland India." *Report on Ecological and Social Impact Assessments Post-Tsunami in Mainland India*, 105–49.

Mirza, Shireen. 2019. "Becoming Waste: Three Moments in the Life of Landfills in Mumbai City." *Economic and Political Weekly* 54 (47): 37.

Mitchell, Timothy. 2013. *Rule of Experts: Egypt, Techno-Politics, Modernity.* Berkeley: University of California Press.

Mitchell, William J. T. 2004. *Me++: The Cyborg Self and the Networked City.* Cambridge, MA: MIT Press.

Mitter, Partha. 1994. *Art and Nationalism in Colonial India, 1850–1922: Occidental Orientations.* Cambridge: Cambridge University Press.

MoEF (Ministry of Environment and Forests). 2016. *Solid Waste Management Rules.*

Moore, Amelia. 2012. "The Aquatic Invaders: Marine Management Figuring Fishermen, Fisheries, and Lionfish in the Bahamas." *Cultural Anthropology* 27 (4): 667–88.

Moore, Amelia. 2019. *Destination Anthropocene: Science and Tourism in the Bahamas.* Berkeley: University of California Press.

Mosse, David. 2003. "The Making and Marketing of Participatory Development." In *A Moral Critique of Development: In Search of Global Responsibilities,* edited by Anta Kumar Giri, Philip Quarles van Ufford, 57–89. London: Routledge.

MPCB (Maharashtra Pollution Control Board). 2019. "Annual Report 2018–19."

Mukherjee, Nibedita, Farid Dahdouh-Guebas, Vena Kapoor, Rohan Arthur, Nico Koedam, Aarthi Sridhar, and Kartik Shanker. 2010. "From Bathymetry to Bioshields: A Review of Post-Tsunami Ecological Research in India and Its Implications for Policy." *Environmental Management* 46 (3): 329–39. DOI: 10.1007/s00267-010-9523-1.

Mukhija, Vinit. 2003. *Squatters as Developers? Slum Redevelopment in Mumbai.* New York: Routledge.

Mumbai Live. 2020. "Aaditya Thackeray Launches Two Projects Aimed At Cleaning Mithi River." September 4.

Mumbai Metropolitan Regional Development Authority. 2006 "Development of Action Plan for the Environmental Improvement of the Mithi River and Along Its Banks." Mumbai: CESE IIT Bombay, June.

Municipal Corporation of Greater Mumbai. "Fact Finding Committee on the Mumbai Floods." 2006. Mumbai.

Münster, Daniel, and Ursula Münster. 2012. "Consuming the Forest in an Environment of Crisis: Nature Tourism, Forest Conservation and Neoliberal Agriculture in South India." *Development and Change* 43 (1): 205–27. DOI: 10.1111/j.1467-7660.2012.01754.x.

Murphy, Michelle. 2013. "Distributed Reproduction, Chemical Violence, and Latency." *Scholar and Feminist Online* 11 (3): 1–7.

Murray, Martin J. 2009. "Fire and Ice: Unnatural Disasters and the Disposable Urban Poor in Post-Apartheid Johannesburg." *International Journal of Urban and Regional Research* 33 (1): 165–92. DOI: 10.1111/j.1468-2427.2009.00835.x.

Nair, Gayatri. 2021. *Set Adrift: Capitalist Transformations and Community Politics along Mumbai's Shores*. Oxford: Oxford University Press.

Nair, Shalini. 2019. "Gases Spook Comps in IT Park Built on Dump | Mumbai News—Times of India." *Times of India*, February 4, 2019. https://timesofindia. indiatimes.com/city/mumbai/Gases-spook-comps-in-IT-park-built-on-dump/articleshow/1842135.cms.

Nallathiga, Ramakrishna. 2009. "From Master Plan to Vision Plan: The Changing Role of Plans and Plan Making in City Development (with Reference to Mumbai)." *Theoretical and Empirical Researches in Urban Management* 4 (13): 141–57. https://www.jstor.org/stable/24872623.

Narayanan, Nayantara. 2014. "Sewage from Mumbai and Karachi Is Killing Fisheries in the Arabian Sea." *Scroll*, October 14. https://scroll.in/article/683461/sewage-from-mumbai-and-karachi-is-killing-fisheries-in-the-arabian-sea.

National Centre For Earth Science Studies. 2014. "Coastal Zone Management Plan Of Kochi Corporation." Kochi, Kerala: Kerala Government.

"National Disaster Management Guidelines: Management of Urban Flooding." 2010. National Disaster Management Authority, Government of India, September.

Nayak, Shailesh. 2002. "Use of Satellite Data in Coastal Mapping." *Indian Cartographer* 22: 147–57.

Nayak, Shailesh. 2004. "Application of Remote Sensing for Implementation of Coastal Zone Regulations: A Case Study of India." In *Proceedings of the 7th Conference on Global Data Infrastructure, 2004*, 2–6. Bengaluru, Karnataka: Global Spatial Data Infrastructure. http://gsdidocs.org/gsdiconf/GSDI-7/papers/TSlpSN.pdf.

Nijman, Jan. 2008. "Against the Odds: Slum Rehabilitation in Neoliberal Mumbai." *Cities* 25 (2): 73–85.

Nixon, Rob. 2011. *Slow Violence and the Environmentalism of the Poor*. Cambridge, MA: Harvard University Press.

November, Valérie, Eduardo Camacho-Hübner, and Bruno Latour. 2010. "Entering a Risky Territory: Space in the Age of Digital Navigation." *Environment and Planning D: Society and Space* 28 (4): 581–99. DOI: 10.1068/d10409.

Olwig, M. F., M. K. Sørensen, M. S. Rasmussen, F. Danielsen, V. Selvam, L. B. Hansen, L. Nyborg, K. B. Vestergaard, F. Parish, and V. M. Karunagaran. 2007. "Using Remote Sensing to Assess the Protective Role of Coastal Woody Vegetation Against Tsunami Waves." *International Journal of Remote Sensing* 28 (13–14): 3153–69. DOI: 0.1080/01431160701420597.

"1-D Mathematical Model and Desk Studies for Mitigating Floods of Mithi River in Mumbai." 2006. Central Water and Power Research Station, Mumbai.

Orlove, Benjamin S. 1991. "Mapping Reeds and Reading Maps: The Politics of Representation in Lake Titicaca." *American Ethnologist* 18 (1): 3–38. https://www.jstor.org/stable/645563.

Ota, Yoshitaka, and Roger Just. 2008. "Fleet Sizes, Fishing Effort, and the 'Hidden' Factors Behind Statistics: An Anthropological Study of Small-Scale Fisheries in UK." *Marine Policy* 32 (3): 301–8. DOI: 10.1016/j.marpol.2007.06.006.

Packel, Dan. 2011. "Snacking with the Sons of the Soil." *Gastronomica: The Journal of Food and Culture* 11 (1): 67–70. DOI: 10.1525/gfc.2011.11.1.67.

Padawangi, Rita, Etienne Turpin, Michaela F. Prescott, Ivana Lee, and Ariel Shepherd. 2016. "Mapping an Alternative Community River: The Case of the Ciliwung." *Sustainable Cities and Society* 20: 147–57. DOI: 10.1016/j.scs.2015.09.001.

Pandian, Anand. 2019. *A Possible Anthropology*. Durham, NC: Duke University Press.

Parappurathu, Shinoj, and C. Ramachandran. 2017. "Taming the Fishing Blues: Reforming the Marine Fishery Regulatory Regime in India." *Economic and Political Weekly* 52 (45): 73–81. https://www.jstor.org/stable/26697845.

Pardeshi, Peehu, Balaram Jadhav, Ravikant Singh, Namrata Kapoor, Ronita Bardhan, Arnab Jana, Siddarth David, and Nobhojit Roy. 2020. "Association between Architectural Parameters and Burden of Tuberculosis in Three Resettlement Colonies of M-East Ward, Mumbai, India." *Cities and Health*, 1–18. DOI: 10.1080/23748834.2020.1731919.

Parker, Brenda. 2006. "Constructing Community Through Maps? Power and Praxis in Community Mapping." *The Professional Geographer* 58 (4): 470–84. DOI: 10.1111/j.1467-9272.2006.00583.x

Parthasarathy, D. 2011. "Hunters, Gatherers, and Foragers in a Metropolis: Commonising the Private and Public in Mumbai." *Economic and Political Weekly* 46 (50): 55. https://www.jstor.org/stable/41319484.

Parthasarathy, Devanathan. 2015. "Informality, Resilience, and the Political Implications of Disaster Governance." *Pacific Affairs* 88 (3): 551–75. DOI: 10.5509/2015883551.

Patel, Sheela. 2004. "Tools and Methods for Empowerment Developed by Slum and Pavement Dwellers' Federations in India." *Participatory Learning and Action* 50: 117–30.

Patel, Sheela, and Jockin Arputham. 2008. "Plans for Dharavi: Negotiating a Reconciliation between a State-Driven Market Redevelopment and Residents' Aspirations." *Environment and Urbanization* 20 (1): 243–53. DOI: 10.1177/0956247808089161.

Patel, Sheela, Celine d'Cruz, and Sundar Burra. 2002. "Beyond Evictions in a Global City: People-Managed Resettlement in Mumbai." *Environment and Urbanization* 14 (1): 159–72. DOI: 10.1177/095624780201400113.

Pathak, Gauri. 2020. "Permeable Persons and Plastic Packaging in India: From Biomoral Substance Exchange to Chemotoxic Transmission." *Journal of the Royal Anthropological Institute* 26 (4): 751–65. DOI: 10.1111/1467-9655.13365.

Patil, Supriya S., and Arvind D. Shaligram. 2013. "Refractometric Fiber Optic Sensor for Detecting Salinity of Water." *Journal of Sensor Technology* 3 (3): 70. DOI: 10.4236/jst.2013.33012.

Paul, Anirudh, Prasad Shetty, and Shekhar Krishnan. 2005. "The City as Extracurricular Space: Re-Instituting Urban Pedagogy in South Asia." *Inter-Asia Cultural Studies* 6 (3): 386–409. DOI: 0.1080/14649370500170027.

Peke, Shuddhawati. 2013. "Women Fish Vendors in Mumbai: A Study Report." Monograph 978-93-80802-13-8. Chennai: International Collective in Support of Fishworkers.

Peluso, Nancy Lee. 1995. "Whose Woods Are These? Counter-Mapping Forest Territories in Kalimantan, Indonesia." *Antipode* 27 (4): 383–406. DOI: doi. org/10.1111/j.1467-8330.1995.tb00286.x.

Phatak, Vidyadhar K., and Shirish B. Patel. 2005. "Would Decentralisation Have Made a Difference?" *Economic and Political Weekly* 40 (36): 3902–5. https://www.jstor. org/stable/4417100.

Pinho, Patricia, Ben Orlove, and Mark Lubell. 2012. "Overcoming Barriers to Collective Action in Community-Based Fisheries Management in the Amazon." *Human Organization* 71 (1): 99–109. DOI: 10.17730/ humo.71.1.c34057171x0w8g5p.

Pinney, Christopher. 1997. *Camera Indica: The Social Life of Indian Photographs.* Chicago: University of Chicago Press.

Pitchon, Ana. 2015. "Large-Scale Aquaculture and Coastal Resource-Dependent Communities: Tradition in Transition on Chiloe Island, Chile." *Journal of Latin American and Caribbean Anthropology* 20 (2): 343–58. DOI: 10.1111/jlca.12151.

Poole, Deborah. 1997. *Vision, Race, and Modernity: A Visual Economy of the Andean Image World.* Vol. 13. Princeton, NJ: Princeton University Press.

Prakash, Gyan. 2010. *Mumbai Fables.* Princeton, NJ: Princeton University Press.

Probyn, Elspeth. 2016. *Eating the Ocean.* Durham, NC: Duke University Press.

Punekar, Vijaya B. 1959. *The Son Kolis of Bombay.* Mumbai: Popular Book Depot.

Rademacher, Anne. 2011. *Reigning the River.* Durham, NC: Duke University Press.

Rafiq, Farhat, Sirajuddin Ahmed, Shamshad Ahmad, and Amir Ali Khan. 2016. "Urban Floods in India." *International Journal of Scientific and Engineering Research* 7: 721–34. ISSN 2229-5518.

Raj, Kapil. 2007. *Relocating Modern Science: Circulation and the Construction of Knowledge in South Asia and Europe, 1650–1900.* New York: Springer.

Raje, S. G., and Vinay D. Deshmukh. 1989. "On the Dol Net Operation at Versova, Bombay." *Indian Journal of Fisheries* 36 (3): 239–48.

Raman, Bhavani. 2012. *Document Raj: Writing and Scribes in Early Colonial South India.* Chicago: University of Chicago Press.

Ramaswamy, Sumathi. 2014. "Introduction: The Work of Vision in the Age of European Empires." In *Empires of Vision*, edited by Sumathi Ramaswamy and Martin Jay, 1–22. Durham, NC: Duke University Press.

Ramesh, Aditya. 2021. "Flows and Fixes: Water, Disease and Housing in Bangalore, 1860–1915." *Urban History*, 1–23. DOI: 10.1017/S0963926821000705.

Randeria, Shalini. 2003. "Glocalization of Law: Environmental Justice, World Bank, NGOs, and the Cunning State in India." *Current Sociology* 51 (3–4): 305–28. DOI: 10.1177/0011392103051003000.

Ranganathan, Malini. 2015. "Storm Drains as Assemblages: The Political Ecology of Flood Risk in Post-Colonial Bangalore." *Antipode* 47 (5): 1300–1320. DOI: 10.1111/anti.12149.

Rao, Divya Badami, and M. V. Ramana. 2008. "Violating Letter and Spirit: Environmental Clearances for Koodankulam Reactors." *Economic and Political Weekly* 51 (43): 14–18. https://www.jstor.org/stable/40278304.

Rao, Vyjayanthi. 2007. "Proximate Distances: The Phenomenology of Density in Mumbai." *Built Environment* 33 (2): 227–48. https://www.jstor.org/stable/23289578.

Reddy, Rajyashree N. 2015. "Producing Abjection: E-Waste Improvement Schemes and Informal Recyclers of Bangalore." *Geoforum* 62: 166–74. DOI: 10.1016/j.geoforum.2015.04.003.

Reeves, Peter, Andrew Pope, John McGuire, and Bob Pokrant. 1996. "The Koli and the British at Bombay: The Structure of Their Relations to the Mid-Nineteenth Century." *South Asia: Journal of South Asian Studies* 19 (s1): 97–119. DOI: 10.1080/00856409608723274.

Reno, Joshua. 2016. *Waste Away: Working and Living with a North American Landfill.* Berkeley: University of California Press.

Roxy, M. K., Subimal Ghosh, Amey Pathak, R. Athulya, Milind Mujumdar, Raghu Murtugudde, Pascal Terray, and M. Rajeevan. 2017. "A Threefold Rise in Widespread Extreme Rain Events Over Central India." *Nature Communications* 8 (1) (October 3): 1–11. DOI: 10.1038/s41467-017-00744-9.

Roxy, Mathew Koll, Aditi Modi, Raghu Murtugudde, Vinu Valsala, Swapna Panickal, S. Prasanna Kumar, M. Ravichandran, Marcello Vichi, and Marina Lévy. 2016. "A Reduction in Marine Primary Productivity Driven by Rapid Warming over the Tropical Indian Ocean." *Geophysical Research Letters* 43 (2): 826–33. DOI: 10.1002/2015GL066979.

Roy, Ananya. 2005. "Urban Informality: Toward an Epistemology of Planning." *Journal of the American Planning Association* 71 (2): 147–58. DOI: 10.1080/01944360508976689.

Roy, Saumya. 2022. "Rain, Rumours And Remorse." *Outlook*, February 5. https://www. outlookindia.com/magazine/story/rain-rumours-and-remorse/228282.

Rushdie, Salman. 2010. *Midnight's Children*. New York: Random House.

Sahu, Amiya Kumar. 2007. "Present Scenario of Municipal Solid Waste (MSW) Dumping Grounds in India." In *Proceedings of the International Conference on Sustainable Solid Waste Management*, Mumbai, India, 5–7.

Sahu, Basanta Kumar. 2016. "Genesis of Coastal Regulations in India." *ENVIS Centre of Odisha's State of Environment* (blog). 2016. http://orienvis.nic.in/indexx.aspx?la ngid=1&slid=651&mid=2&sublinkid=189.

Salagrama, Venkatesh. 2006. "Trends in Poverty and Livelihoods in Coastal Fishing Communities of Orissa State, India." FAO Fisheries Technical Paper 490. Rome: FAO.

Salve, Pradeep S. 2020. "A Comparative Study of Prevalence of Morbidities among Municipal Solid Waste Workers in Mumbai." *SN Comprehensive Clinical Medicine* 2: 1534–42.

Salve, Pradeep S., Praveen Chokhandre, and Dhananjay W. Bansod. 2019. "Substance Use among Municipal Solid Waste Workers in Mumbai: A Cross-Sectional Comparative Study." *Journal of Substance Use* 24 (4): 432–38. DOI: 10.1007/ s42399-020-00441-7.

Sanyal, Bishwapriya, and Vinit Mukhija. 2001. "Institutional Pluralism and Housing Delivery: A Case of Unforeseen Conflicts in Mumbai, India." *World Development* 29 (12): 2043–57. DOI: 10.1016/S0305-750X(01)00082-1.

Schneider, Birgit, and Thomas Nocke. 2014. "Image Politics of Climate Change: Introduction." In *Image Politics of Climate Change: Visualizations, Imaginations, Documentations*, edited by Birgit Schneider and Thomas Nocke, 9–29. Cambridge, UK: Cambridge University Press.

Schuster, Caroline E. 2023. *Forecasts: A Story of Weather and Finance at the Edge of Disaster*. Toronto: University of Toronto Press.

Schwenkel, Christina. 2015. "Spectacular Infrastructure and Its Breakdown in Socialist Vietnam." *American Ethnologist* 42 (3): 520–34. DOI: 10.1111/amet.12145.

Scott, James C. 1998. *Seeing like a State: How Certain Schemes to Improve the Human Condition Have Failed*. New Haven, CT: Yale University Press.

Searle, Llerena Guiu. 2016. *Landscapes of Accumulation: Real Estate and the Neoliberal Imagination in Contemporary India*. Chicago: University of Chicago Press.

Sehara, D. B. S., and J. P. Kharbari. 1987. "Study on 'Dol' Net Fishery at Selected Centres in Northwest Coast with Special Reference to Costs and Returns." Marine Fisheries Information Service, Technical & Extension Series 78: 1–15.

Sharma, Chandrika. 1996. "Coastal Area Management in South Asia: A Comparative Perspective." Background Paper: South Asia Workshop on Fisheries and Coastal Area Management. Chennai, Tamil Nadu: International Collective in Support of Fishworkers. http://aquaticcommons.org/277/.

Shaw, Annapurna. 2004. *The Making of Navi Mumbai*. Hyderabad: Orient Blackswan.

"Shoreline Change Atlas of India (Volume—2 Maharashtra and Goa)." 2014. Atlas SAC/EPSA/GSAG/GSD/A/01/14. New Delhi: Space Applications Centre (ISRO) and Coastal Erosion Directorate, Central Water Commission, Ministry of Water Resources.

Singh, Diyesh. 2011. "How Pavit Breached 3-Tier Security." *DNA*, August 1, 2011, Mumbai edition. https://www.dnaindia.com/mumbai/report-how-pavit-breached-3-tier-security-1571793.

Siriwardane-de Zoysa, Rapti, and Anna-Katharina Hornidge. 2022. "Tidal Turns: Coastal Urbanities in Island Southeast Asia." In *Coastal Urbanities: Mobilities, Meanings, Manoeuvrings*, edited by Rapti Siriwardane-de Zoysa, Kelvin E. Y. Low, Noorman Abdullah, and Anna-Katharina Hornidge, 1–23. Brill. DOI: 10.1163/9789004523340_002.

Sivaramakrishnan, K. 2011. "Environment, Law, and Democracy in India." *Journal of Asian Studies* 70 (04): 905–28.

Sopranzetti, Claudio, Sara Fabbri, and Chiara Natalucci. 2021. *The King of Bangkok*. Toronto: University of Toronto Press.

Sousanis, Nick. 2015. *Unflattening*. Cambridge, MA: Harvard University Press.

Sridhar, Aarthi. 2006. "Pre-and Post-Tsunami Coastal Planning and Land-Use Policies and Issues in India." Food and Agriculture Organization. Rome: FAO. http://www.fao.org/forestry/13139-053ffbd98f6f91d58a5dd2e4046fdc152.pdf.

Sriraman, Tarangini. 2018. *In Pursuit of Proof: A History of Identification Documents in India*. New Delhi: Oxford University Press.

Stamatopoulou-Robbins, Sophia. 2014. "Occupational Hazards." *Comparative Studies of South Asia, Africa and the Middle East* 34 (3): 476–96. DOI: 10.1215/1089201X-2826049.

Star, Susan Leigh. 2010. "This Is Not a Boundary Object: Reflections on the Origin of a Concept." *Science, Technology, and Human Values* 35 (5): 601–17. DOI: 10.1177/0162243910377624.

Strip, Percival, and Olivia Strip. 1944. *The Peoples of Bombay*. Bombay: Thacker.

Subramanian, Ajantha. 2009. *Shorelines: Space and Rights in South India*. 1st ed. Stanford, CA: Stanford University Press.

Sundar, Aparna. 2014. "From Regulation to Management and Back Again: Exploring Governance Shifts in India's Coastal Zone." *Conservation and Society* 12 (4): 364–75. https://www.jstor.org/stable/26393171.

Swaminathan Research Foundation. 2005. "Report of the Expert Committee on Coastal Regulation Zone Notification, 1991." New Delhi: Ministry of Environment, Forests, and Climate Change, http://iomenvis.in/pdf_documents/MSS_Report.pdf.

Swaminathan, M. S., Shailesh Nayak, Sunita Narain, and J. M. Mauskar. 2009. "Final Frontier: Agenda to Protect the Ecosystem and Habitat of India's Coast for Conservation and Livelihood Security." Expert Committee on the Draft Coastal Management Zone (CMZ) Notification. New Delhi: Ministry of Environment, Forests, and Climate Change. http://www.indiaenvironmentportal.org.in/files/cmz_report.pdf.

Sze, Julie, and Gerardo Gambirazzio. 2013. "Eco-Cities without Ecology: Constructing Ideologies, Valuing Nature." In *Resilience in Ecology and Urban Design*, edited by S.T.A. Pickett, M.L. Cadenasso, Brian McGrath, 289–97. New York: Springer. DOI: 10.1007/978-94-007-5341-9_14.

Taussig, Michael. 2011. *I Swear I Saw This: Drawings in Fieldwork Notebooks, Namely My Own*. Chicago: University of Chicago Press.

Ten Bos, René. 2009. "Towards an Amphibious Anthropology: Water and Peter Sloterdijk." *Environment and Planning D: Society and Space* 27 (1): 73–86. DOI: 10.1068/d13607.

Theodossopoulos, Dimitrios. 2022. "Introduction: Graphic Ethnography on the Rise." Theorizing the Contemporary, Fieldsights, July 28. https://culanth.org/fieldsights/series/graphic-ethnography-on-the-rise.

Thongchai, Winichakul. 1994. *Siam Mapped: A History of the Geo-Body of a Nation.* Honolulu: University of Hawaii Press.

Times News Network. 2005. "A Cloudburst Over Mumbai?" *Economic Times,* July 29. https://economictimes.indiatimes.com/a-cloudburst-over-mumbai/articleshow/1185043.cms?from=mdr.

Times News Network. 2019. "Monsoon Covers Ground: Mumbai Gets 97% of June Rain in Two Days." *Times of India,* June 30. https://timesofindia.indiatimes.com/city/mumbai/monsoon-covers-ground-mumbai-gets-97-of-june-rain-in-two-days/articleshow/70006959.cms.

TISS. 2015. "Socioeconomic Conditions and Vulnerabilities: A Report of the Baseline Survey of M (East) Ward of Mumbai." Tata Institute of Social Sciences, Mumbai.

Todd, Zoe. 2014. "Fish Pluralities: Human-Animal Relations and Sites of Engagement in Paulatuuq, Arctic Canada." *Études/Inuit/Studies* 38 (1–2): 217–38. DOI: 10.7202/1028861ar.

Tsing, Anna Lowenhaupt. 2015. *The Mushroom at the End of the World: On the Possibility of Life in Capitalist Ruins.* Princeton, NJ: Princeton University Press.

Urla, Jacqueline. 1993. "Cultural Politics in an Age of Statistics: Numbers, Nations, and the Making of Basque Identity." *American Ethnologist* 20 (4): 818–43. DOI: 10.1525/ae.1993.20.4.02a00080.

Vaidya, S. S., and J. R. Kulkarni. 2007. "Simulation of Heavy Precipitation over Santacruz, Mumbai on 26 July 2005, Using Mesoscale Model." *Meteorology and Atmospheric Physics* 98 (1) (October 1): 55–66. DOI: 10.1007/s00703-006-0233-4.

van Dooren, Thom. 2014. *Flight Ways: Life and Loss at the Edge of Extinction.* New York: Columbia University Press.

Venkataramani, Chitra. 2017. "Identification, 12 materiality and housing transformations in Mumbai." *Trends and Issues in Housing in Asia: Coming of an Age,* edited by Annapurna Shaw and Urmi Sengupta, 278–300. London: Routledge.

Verma, Ashok, S. Balachandran, Naresh Chaturvedi, and Vinod Patil. 2004. "A Preliminary Report on the Biodiversity of Mahul Creek, Mumbai, India with Special Reference to Avifauna." *Zoos' Print Journal* 19 (9): 1599–1605. DOI: 10.11609/JoTT.ZPJ.1172.1599-605.

Vijay, V., Sanjay Biradar, Inamdar Arun, and Geetanjali Deshmukhe. 2005. "Mangrove Mapping and Change Detection around Mumbai (Bombay) Using Remotely Sensed Data." *Indian Journal of Marine Sciences* 34 (3): 310–15.

Vivekanandan, E., V. V. Singh, and Joe K. Kizhakudan. 2013. "Carbon Footprint by Marine Fishing Boats of India." *Current Science,* 361–66. https://www.jstor.org/stable/24097968.

Volkman, Toby Alice. 1994. "Our Garden Is the Sea: Contingency and Improvisation in Mandar Women's Work." *American Ethnologist* 21 (3): 564–85. DOI: 10.1525/ae.1994.21.3.02a00060.

Von Schnitzler, Antina. 2017. *Democracy's Infrastructure: Techno-Politics and Protest after Apartheid.* Princeton, NJ: Princeton University Press.

Vyas, V. S., and V. Ratna Reddy. 1998. "Assessment of Environmental Policies and Policy Implementation in India." *Economic and Political Weekly* 33 (1/2): 48–54.

Wagh, Shweta, and Hussain Indorewala. 2019. "Why the Bombay High Court Decisively Stopped the Coastal Road Project." *Scroll,* July 25. https://science.thewire.in/law/bombay-high-court-coastal-road-project/.

Wainwright, Joel, and Joe Bryan. 2009. "Cartography, Territory, Property: Postcolonial Reflections on Indigenous Counter-Mapping in Nicaragua and Belize." *Cultural Geographies* 16 (2): 153–78. DOI: 10.1177/1474474008101515.

Warhaft, Sally. 2001. "No Parking at the Bunder: Fisher People and Survival in Capitalist Mumbai." *South Asia: Journal of South Asian Studies* 24 (2): 213–23. DOI: 10.1080/00856400108723458.

Weinstein, Liza, Andrew Rumbach, and Saumitra Sinha. 2019. "Resilient Growth: Fantasy Plans and Unplanned Developments in India's Flood-Prone Coastal Cities." *International Journal of Urban and Regional Research* 43 (2): 273–91. DOI: 10.1111/1468-2427.12743.

Whitington, Jerome. 2016. "Carbon as a Metric of the Human." *PoLAR: Political and Legal Anthropology Review* 39 (1): 46–63. DOI: 10.1111/plar.12130.

Wiber, Melanie Gay, Sheena Young, and Lisette Wilson. 2012. "Impact of Aquaculture on Commercial Fisheries: Fishermen's Local Ecological Knowledge." *Human Ecology* 40 (1): 29–40. DOI: 0.1007/s10745-011-9450-7.

Wilson, James A., James M. Acheson, Mark Metcalfe, and Peter Kleban. 1994. "Chaos, Complexity, and Community Management of Fisheries." *Marine Policy* 18 (4): 291–305. DOI: 10.1016/0308-597X(94)90044-2.

Yu, Lily E. 2011. "The Cartographer Wasps and the Anarchist Bees." *Clarksworld Magazine* 55 (April).

Zhao Nan, Shen Junli, Jia Lijing, 2009. "An Empirical Study on Infrastructure Carrying Capacity and Carrying Condition of Beijing." *Urban Studies* 16 (4): 68–75.